Proceedings

of the

3rd Beilstein Glyco-Bioinformatics Symposium

Discovering the Subtleties of Sugars

June 10th – 14th, 2013

Potsdam, Germany

Edited by Martin G. Hicks and Carsten Kettner

Beilstein-Institut zur Förderung der Chemischen Wissenschaften

Trakehner Str. 7 – 9
60487 Frankfurt
Germany

Telephone: +49 (0)69 7167 3211
Fax: +49 (0)69 7167 3219
E-Mail: info@beilstein-institut.de
Web-Page: www.beilstein-institut.de

Impressum

Discovering the Subtleties of Sugars, Martin G. Hicks and Carsten Kettner (Eds.), Proceedings of the Beilstein Glyco-Bioinformatics Symposium, June 10th – 14th 2013, Potsdam, Germany.

Bibliographic information published by the *Deutsche Nationalbibliothek*. The *Deutsche Nationalbibliothek* lists this publication in the *Deutsche Nationalbibliografie*; detailed bibliographic data are available in the Internet at http://dnb.ddb.de.

ISBN 978-3-8325-3948-1

Layout by: Hübner Electronic Publishing GmbH
Steinheimer Straße 22a
65343 Eltville
www.huebner-ep.de

Printed by: Logos Verlag Berlin GmbH
Comeniushof, Gubener Str. 47
10243 Berlin
www.logos-verlag.de

Cover Illustration by: Bosse und Meinhard
Kaiserstraße 34
53113 Bonn
www.bosse-meinhard.de

PREFACE

The Beilstein Symposia address contemporary issues in the chemical and related sciences by employing and interdisciplinary approach. Scientists from a wide range of areas are invited to present aspects of their work for discussion, with the aim of not only advancing sciences, but also enhancing interdisciplinary communication. Traditionally, the Beilstein Symposia are kept small with up to 50 participants to provide a convivial atmosphere for the both lectures and lively discussions and the ready exchange of thoughts and ideas.

The appreciation and understanding of the role that carbohydrates play in nature has grown over the last few years driven by the advances in our ability to analyze and synthesize their structures. Their role not only as primary energy-storage molecules but also as structural modifiers of e.g. glycoproteins and glycolipids, as well as in physiological and pathological events such as adherence, cell-cell interaction, transport, signaling and protection is becoming clearer and more accessible to researchers. Over the last decade the fields of glycobiology and glycochemistry in combination with in-silico applications have been augmented by a further field – glycomics. A major aim of glycomics research is to achieve a comprehensive identification and characterization of the repertoire of glycan structures present in an organism, cell or tissue at a defined time. The continual improvement of analysis methods and computational techniques leads to glycan characterization and identification with increased depth, speed and efficiency but also generates ever increasing amounts of data of variable quality and completeness.

Thus the many web-accessible repositories result in a highly fragmented knowledgebase which in consequence complicates the development and application of bioinformatics tools for the analysis of this data.

This situation has led to a general consensus that community wide efforts should be spent towards consolidating and systematizing the collective knowledgebase with integration of universal bioinformatics tools for both the representation, mining as well as annotation of experimental data sets to advance and interface glycomics with related genomics and proteomics projects. Additionally, both experimentalists and bioinformaticians also expressed their demands for data reporting practices that include the comprehensive description of conditions, techniques and experimental results to enable researchers to evaluate the degree of structural definitions, to interpret the results and to reproduce the experiments.

The previous symposia held in 2009 and 2011 brought the stakeholders in the area of glycomics together and provided a platform to discuss the role of bioinformatics in this emerging field. One important outcome was the founding of a new working group called MIRAGE (Minimum Information Required for a Glycomics Experiment) under the auspices of the Beilstein-Institut. This group has the function, with involvement of the scientific community, to draw up proposals for reporting standards for glycomics experiments and for setting up a framework to integrate glyco-bioinformatics in a comprehensive platform that will serve biologists, chemists and all interested in glycosciences.

This symposium continued successfully to bring together those scientists that "produce" data with those that "use" the data and make it available to the community. In particular, in their presentations speakers delivered insights into the diverse physiological and structural subtleties of sugars by covering aspects such as: structure-function relationships of carbohydrates, modeling carbohydrate structure and carbohydrate interactions with other biomacromolecules, deciphering carbohydrate signals, carbohydrate identification, annotation and analysis, metadata for the description of glycomics experiments, software tools for data mining and analysis.

We would like to thank particularly the authors who provided us with written versions of the papers that they presented. Special thanks go to all those involved with the preparation and organization of the symposium, to the chairmen who piloted us successfully through the sessions and to the speakers and participants for their contribution in making this symposium a success.

Frankfurt/Main, December 2014

Martin G. Hicks
Carsten Kettner
Peter Seeberger

Contents

Page

Metabolic Engineering of Bacteria

Elisabeth Memmel and Jürgen Seibel*

Institute of Organic Chemistry, University of Würzburg, Am Hubland,
97074 Würzburg, Germany

E-Mail: *seibel@chemie.uni-wuerzburg.de

Received: 3rd December 2013 / Published: 22nd December 2014

Abstract

Metabolic glycoengineering is a technique introduced in the early 90s of the last century by Reutter *et al.*. It utilises the ability of cells to metabolically convert sugar derivatives with bioorthogonal side chains like azides or alkynes and by that incorporation into several glyco structures. Afterwards, the carbohydrates can be labelled to study their distribution, dynamics and roles in different biological processes. So far many studies were performed on mammal cell lines as well as in small animals. Very recently, bacterial glyco-structures were targeted by glycoengineering, showing promising results in infection prevention by reducing pathogen adhesion towards human epithelial cells.

Introduction

Bacteria were among the first life forms to appear on earth, and are present in most habitats on the planet, e. g., they live in symbiosis with plants and animals. Compared to human cells there are ten times as many bacterial cells in our body. Most of them are harmless or even beneficial. But some species are pathogenic and cause infectious diseases with more than 1.2 million deaths each year [1]. Those infections include cholera, syphilis, anthrax, leprosy, and bubonic plague as well as respiratory infections like tuberculosis.

This article is part of the Proceedings of the Beilstein Glyco-Bioinformatics Symposium 2013.
www.proceedings.beilstein-symposia.org

Bacterial Cell Surface Architecture

Bacteria are divided into Gram-positive and Gram-negative species. Gram-positive bacteria are surrounded by a peptidoglycan cell wall. This peptidoglycan is a polymer consisting of alternating β-(1,4) linked *N*-acetylglucosamine and *N*-acetylmuramic acids which are cross-linked by four amino acids (D-alanine, L-lysine, D-glutamine and L-alanine) to form a mesh of 20 – 80 nm diameter (Figure 1).

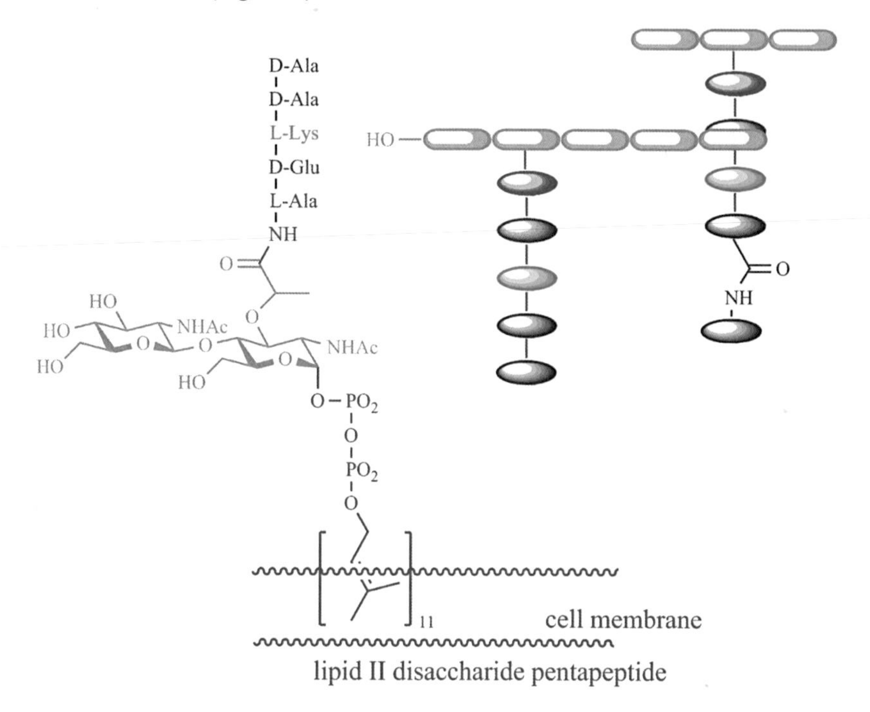

Figure 1. Structure of bacterial peptidoglycan, anchored in the cell membrane.

Gram-negative bacteria are covered by a dense layer of lipopolysaccharides and embedded in their outer membrane. One of the essential components of lipopolysaccharides is 3-deoxy-D-manno-octulosonic acid (KDO).

Metabolic Glycoengineering

Metabolic engineering has demonstrated to be a powerful tool for the modification of eukaryotic cell surfaces. It has been applied in living organisms like human cell cultures, mice and zebrafish [2 – 7]. The principle is based on the incorporation of artificial modified monosaccharides such as derivatives of *N*-acetyl neuraminic acid, *N*-acetylmannosamine, *N*-acetylglucosamine, *N*-acetylgalactosamine and fucose in cell surface glycoproteins [8]. Starting from *N*-acetylglucosamine they are passed through the biosynthetic metabolic pathway and finally transformed to sialic acid [9]. The incorporated modified sugars with functional groups like azides, alkynes and alkenes can bioorthogonally react with their corresponding functionalized probes [8]. Reactions like inverse Electron Demand Diels–Alder reactions of tetrazines and dienophiles such as trans-cyclooctene and cyclopropene, the Sharpless-Huisgen Mendal [3+2] cycloaddition or Staudinger type reaction and photo-crosslinking [10] have been proven to be bioorthogonal (Scheme 1) [11]. In this way, cell surfaces can be further labelled and analysed. However, human cell surfaces differ completely from bacterial ones.

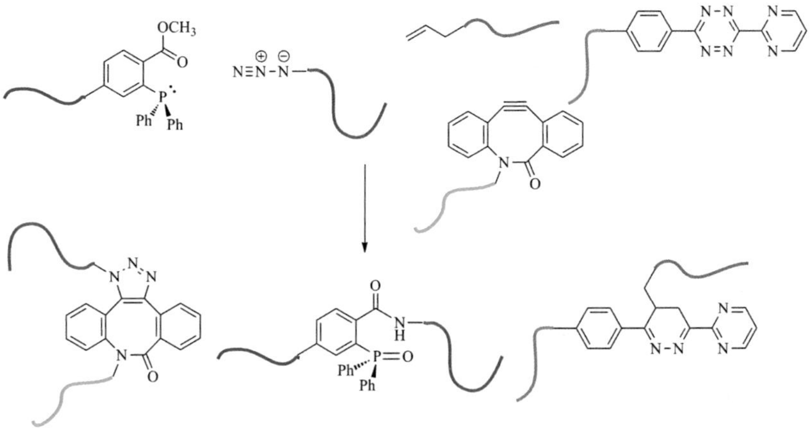

Scheme 1. Established bioorthogonal reactions: Staudinger ligation (blue-red), Huisgen-Sharpless-Mendal cycloaddition (green-red) and the inverse Electron Demand Diels–Alder reaction (iEDDA; purple-orange).

Bacterial cell composition and targets

KDO is a specific component of the inner core of lipopolysaccharides in Gram-negative bacteria. Its biosynthesis starts with arabinose-5-phosphate (arabinose-5-P) which condenses with phosphoenolpyruvate (PEP) to yield KDO-8-phosphate (KDO-8-P). Further dephosphorylation and activation with cytidine monophosphate (CMP) allows the transglycosylation into the lipopolysaccaride core structure [12 – 14].

Dumont *et al.* targeted KDO for the glycoengineering approach in *E. coli* [15]. Referring to the biosynthesis they used 5-azido-5-deoxy-D-arabinofuranose as a precursor for the chemical synthesis of 8-azido-8-deoxy-KDO (Scheme 2). After feeding of *E. coli* with the modified KDO sugar, they were able to stain the bacterial cell surface by azide-alkyne click chemistry with a fluorescent dye (Figure 2) [15]. In general, Gram-positive bacteria such as *Bacillus subtilis* and *Staphylococcus aureus* are missing KDO in their cell wall. Consequently, no labelling was observed under similar conditions.

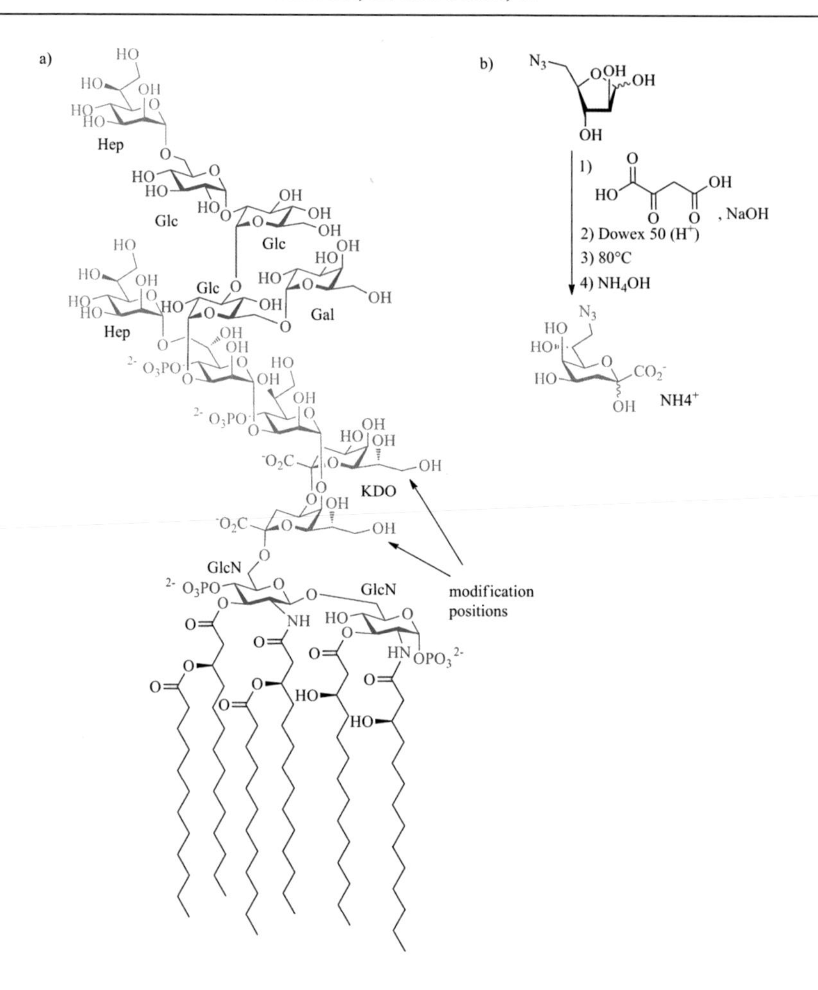

Scheme 2. a) Structure of the major component of *E. coli* K12 lipopolysaccharide. b) Synthesis of N_3-KDO. Gal: D-galactose; Glc: D-glucose; GlcN: 2-amino-2-deoxy-D-glucose; Hep: L-glycero-D-manno-heptose.

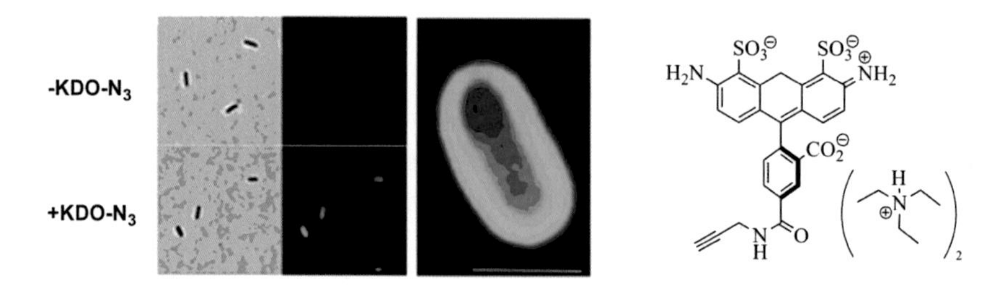

Figure 2. KDO-N_3 metabolically labels *E.coli* lipopolysaccharides. Metabolically incorporated KDO-N_3 in E. coli K12 was revealed by a Cu^I-catalyzed click reaction with alkyne modified fluoresceine (from [15]).

Cell surface labelling and adhesion decrease of **S. aureus**

Bioorthogonal metabolic labelling of Gram-positive *S. aureus* was instead successful when a *N*-acetylglucosamine derivative was used [16]. Bacteria fed with *N*-azidoacetylglucosamine (GlcNAz) could be labelled with alkynylated fluorescent dyes (Tetramethylrhodamine, TAMRA; and Alexa Fluor® 488) using the copper catalysed click reaction (Figure 3). Moreover, by that change the surface properties of *S. aureus*, adhesion towards a human bladder epithelic cell line (T24) could be reduced significantly.

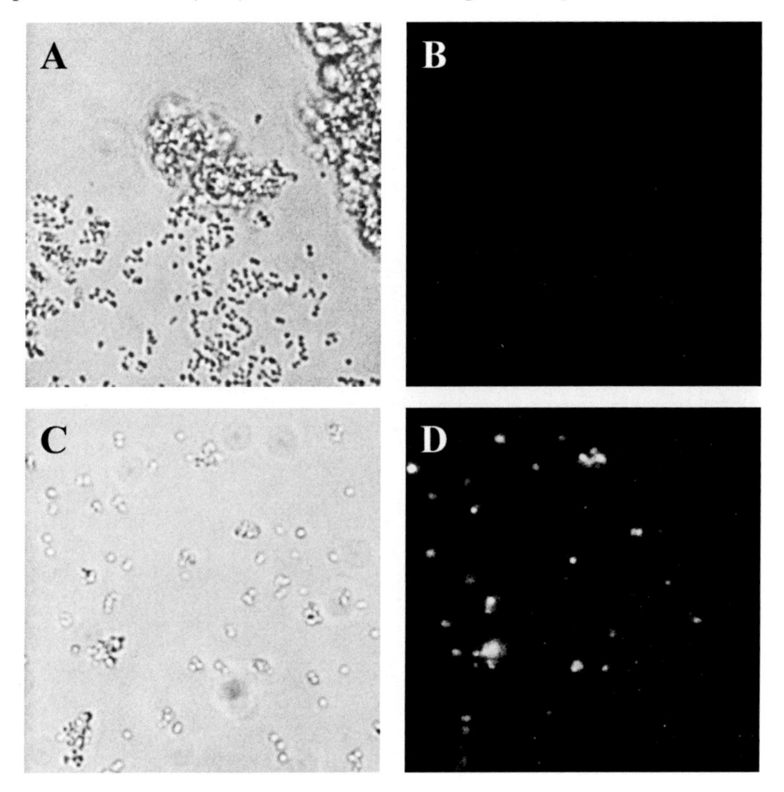

Figure 3. *S. aureus* is labelled with the fluorescent dye Tetramethylrhodamine alkyne by performing a copper catalysed [3+2] cycloaddition on the cell surface after incorporation of the unnatural sugar analogue *N*-azidoacetylglucosamine **(C, D)**. When cultivated in the presence of non-azide containing *N*-acetylglucosamine no fluorescence can be detected **(A, B)**.

Although the exact identity of the targeted glyco structures is still unknown, there are some suggestions. Based on the ability to perform the click reaction on the cell surface as well as the influence on adhesion properties, the affected components may be the peptidoglycan (see Figure 1), glycolipids (teichoic acids) and extracellular polysaccarides (EPS), especially poly-*N*-acetylglucosamine (PNAG) [17]. All of these cell surface components contain potentially accessible GlcNAc as building block (Figure 4).

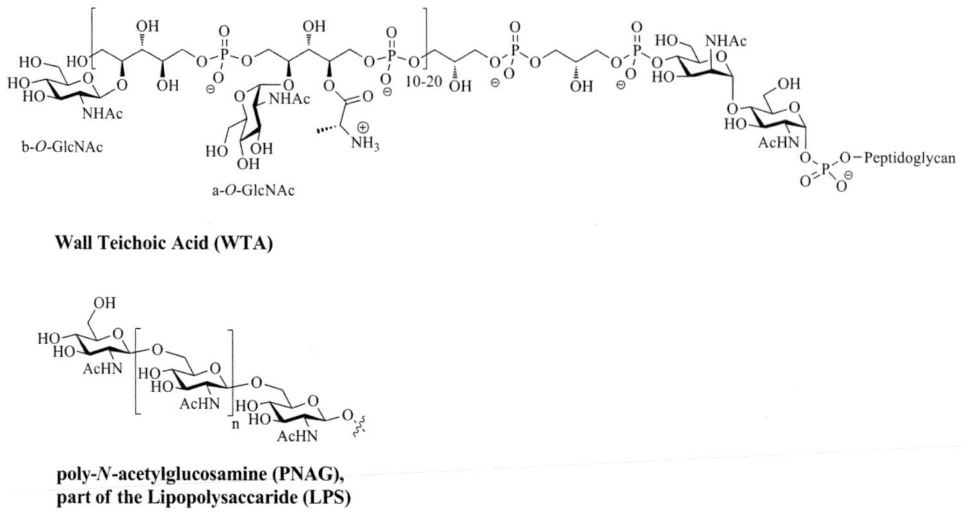

Wall Teichoic Acid (WTA)

poly-*N*-acetylglucosamine (PNAG), part of the Lipopolysaccaride (LPS)

Figure 4. Structures of *S. aureus* cell surface components containing *N*-acetylglucosamine (GlcNAc) as building block.

Furthermore, teichoic acids are discussed to stimulate bacterial adhesion to human epithelial cells [18 – 20]. Glycoengineering of these and other bacterial membrane components might open a new pathway to study, and also to treat bacterial infections.

Perspectives

Complex carbohydrates on cell surfaces play an important role in many biological recognition processes like cell signalling and adhesion as well as in pathogen recognition [21]. While mammal cell surface glycans are investigated extensively, studies on bacterial carbohydrate components and their role in infection processes are still quite rare. To address crucial problems in fighting infections like emerging resistances of bacteria, pathogenesis mechanisms must be understood in more detail. Bacterial adhesion as the first step in an infection process can be influenced or even prevented by changing cell surface properties.

So far, first studies on metabolic engineering of bacterial glycostructures show the possibility to change cell surface properties and influence adhesion as one critical step in infection processes. Novel drugs based on this concept may overcome today's big problems in infection therapy.

Acknowledgements

Elisabeth Memmel thanks the Elite Network of Bavaria scholarship for financial support.

References

[1] WHO http://apps.who.int/gho/data/node.main.CODWORLD? lang = en.

[2] Kayser, H., Zeitler, R., Kannicht, C., Grunow, D., Nuck, R. and Reuter, W. (1992) Biosynthesis of a nonphysiological sialic acid in different rat organs, using *N*-propanoyl-D-hexosamines as precursors. *Journal of Biological Chemistry* **267**:16934 – 16938.

[3] Mahal, L.K., Yarema, K.J., Bertozzi, C.R. (1997) Engineering chemical reactivity on cell surfaces through oligosaccharide biosynthesis. *Science* **276**:1125 – 1128. doi: 10.1126/science.276.5315.1125.

[4] Saxon, E. and Bertozzi, C.R. (2000) Cell surface engineering by a modified Staudinger reaction. *Science* **287**:2007 – 2010. doi: 10.1126/science.287.5460.2007.

[5] Bardor, M., Nguyen, D.H., Diaz, S. and Varki, A. (2005) Mechanism of uptake and incorporation of the non-human sialic acid *N*-glycolylneuraminic acid into human cells. *Journal of Biological Chemistry* **280**:4228 – 4237. doi: 10.1074/jbc.M412040200.

[6] Homann, A., Qamar, R., Serim, S., Dersch, P. and Seibel, J. (2010) Bioorthogonal metabolic glycoengineering of human larynx carcinoma (HEp-2) cells targeting sialic acid. *Beilstein Journal of Organic Chemistry* **6**:24. doi: 10.3762/bjoc.6.24.

[7] Chang, P.V., Prescher, J.A., Sletten, E.M., Baskin, J.M., Miller, I.A., Agard, N. J., Lo, A. and Bertozzi, C.R. (2010) Copper-free click chemistry in living animals. *Proceedings of the National Academy of Sciences of the United States of America* **107**:1821 – 1826. doi: 10.1073/pnas.0911116107.

[8] Laughlin, S.T. and Bertozzi, C.R. (2009) Imaging the glycome. *Proceedings of the National Academy of Sciences of the United States of America* **06**:12 – 17. doi: 10.1073/pnas.0811481106.

[9] Saxon, E., Luchansky, S.J., Hang, H.C., Yu, C., Lee, S.C. and Bertozzi, C.R. (2002) Investigating cellular metabolism of synthetic azidosugars with the Staudinger ligation. *Journal of the American Chemical Society* **124**:14893 – 14902. doi: 10.1021/ja027748x.

[10] Yu, S.-H., Boyce, M., Wands, A.M., Bond, M.R., Bertozzi, C.R. and Kohler, J.J. (2012) Metabolic labeling enables selective photocrosslinking of O-GlcNAc-modified proteins to their binding partners. *Proceedings of the National Academy of Sciences of the United States of America* **109**:4834 – 4839.
doi: 10.1073/pnas.1114356109.

[11] Bertozzi, C.R. and Wu, P. (2013) In vivo chemistry. *Current Opinion in Chemical Biology* **17**:717 – 718.
doi: 10.1016/j.cbpa.2013.10.012.

[12] Lodowska, J., Wolny, D., Weglarz, L. (2013) The sugar 3-deoxy-D-manno-oct-2-ulosonic acid (KDO) as a characteristic component of bacterial endotoxin – a review of its biosynthesis, function, and placement in the lipopolysaccharide core. *Canadian Journal of Microbiology* **59**:645 – 655.
doi: 10.1139/cjm-2013-0490.

[13] Elbein, A.D. and Heath, E.C. (1965) The Biosynthesis of Cell Wall Lipopolysaccharide in *Escherichia Coli*. II. Guanosine Diphosphate 4-Keto-6-Deoxy-D-Mannose, an Intermediate in the Biosynthesis of Guanosine Diphosphate Colitose. *The Journal of Biological Chemistry* **240**:1926 – 1931.

[14] Elbein, A.D. and Heath, E.C (1965) The Biosynthesis of Cell Wall Lipopolysaccharide in *Escherichia Coli*. I. The Biochemical Properties of a Uridine Diphosphate Galactose 4-Epimeraseless Mutant. *The Journal of Biological Chemistry* **240**: 1919 – 1925.

[15] Dumont, A., Malleron, A., Awward, M., Dukan, S. and Vauzeilles, B. (2012) Click-mediated labeling of bacterial membranes through metabolic modification of the lipopolysaccharide inner core. *Angewandte Chemie, International Edition* **51**: 3143 – 3146.
doi: 10.1002/anie.201108127.

[16] Memmel, E., Homann, A., Oelschlaeger, T.A. and Seibel, J. (2013) Metabolic glycoengineering of Staphylococcus aureus reduces its adherence to human T24 bladder carcinoma cells. *Chemical Communications (Cambridge, United Kingdon)* **49**: 7301 – 7303.
doi: 10.1039/c3cc43424a.

[17] Weidenmaier, C. and Peschel, A. (2008) Teichoic acids and related cell-wall glycopolymers in Gram-positive physiology and host interactions. *Nature Reviews Microbiology* **6**:276 – 287.
doi: 10.1038/nrmicro1861.

[18] Livins'ka, O.P., *et al.* (2012) [The influence of teichoic acids from probiotic lactobacilli on microbial adhesion to epithelial cells]. *Mikrobiolohichnyi zhurnal* **74**: 16 – 22.

[19] Weidenmaier, C., Kokai-Kun, J.F., Kristian, S.A., Chanturiya, T., Kalbacher, H., Gross, M., Nicholson, G., Neumeister, B., Mond, J.J. and Peschel, A. (2004) Role of teichoic acids in *Staphylococcus aureus* nasal colonization, a major risk factor in nosocomial infections. *Nature Medicine* **10**:243 – 245.
doi: 10.1038/nm991.

[20] Weidenmaier, C., Peschel, A., Xiong, Y.-Q., Kristian, S.A., Dietz, K., Yeaman, M.R. and Bayer, A.S. (2005) Lack of wall teichoic acids in *Staphylococcus aureus* leads to reduced interactions with endothelial cells and to attenuated virulence in a rabbit model of endocarditis. *The Journal of Infectious Diseases* **191**:1771 – 1777.
doi: 10.1086/429692

[21] Varki, A., *et al.* (2009) in *Essentials of Glycobiology*, 2nd ed. (Eds.: Varki, A., Cummings, R.D. Esko, J.D., Freeze, H.H., Stanley, P., Bertozzi, C.R., Hart, G.W. and Etzler, M. E.), Cold Spring Harbor (NY).

Prediction of Binding Poses and Binding Affinities for Glycans and their Binding Proteins using a Robust Scoring Function for General Protein-Ligand Interactions

Nan-Lan Huang[1] and Jung-Hsin Lin[1,2,3,*]

[1]Research Center for Applied Sciences and [2]Institute of Biomedical Sciences, Academia Sinica, 128 Academia Rd., Sec. 2, Nankang, Taipei 115, Taiwan;

[3]School of Pharmacy, National Taiwan University, 1 Jen-Ai Rd., Sec. 2, Taipei 10051, Taiwan

E-Mail: *jlin@ntu.edu.tw, jhlin@gate.sinica.edu.tw

Received: 30th September 2013 / Published: 22nd December 2014

Abstract

The binding of glycans to proteins represents the major way in which the information contained in glycan structures is recognised, deciphered and put into biological action. The physiological and pathological significance of glycan–protein interactions are drawing increasing attention in the field of structure-based drug design. We have implemented a quantum chemical charge model, the Austin-model 1-bond charge correction (AM1-BCC) method, into a robust scoring function for general protein ligand interactions, called, AutoDock^{RAP}. Here we report its capability to predict the binding poses and binding affinities of glycans to glycan-binding proteins. Our benchmark indicates that this generally applicable scoring function can be adopted in virtual screening of drug candidates and in prediction of ligand binding modes, given the structures of the well-defined recognition domains of glycan-binding proteins.

This article is part of the Proceedings of the Beilstein Glyco-Bioinformatics Symposium 2013.
www.proceedings.beilstein-symposia.org

Introduction

Glycans, the freestanding or covalently linked monosaccharide or oligosaccharide entities in the cells, mediate a wide variety of biological functions of both prokaryotic and eukaryotic cells. A majority of these biological processes involve the recognition of glycans by glycan-binding proteins (GBP). Particularly, glycans located on cell surfaces or secreted biomolecules play a crucial role in cell–cell interaction, including the interaction between host cell and pathogens [1]. For example, understanding of the recognition of host glycoprotein receptors by viral neuraminidase led to the development of high-affinity inhibitors in use to reduce the prevalence of influenza. It is of high priority to understand the molecular basis of the interaction between GBP and glycans involved in the various physiological or pathological events.

Evaluation of the binding affinities of drug-like molecules with the target proteins is crucial for discriminating drug candidates from weak-binding or even nonbinding small molecules. Rigorous statistical mechanical approaches for evaluation of binding free energies are theoretically most satisfactory [2, 3], but such approaches are computationally too demanding for virtual screening. Due to practical consideration, most, if not all, computational docking methods rely greatly on empirical or semi-empirical scoring functions to evaluate protein–ligand interactions. Semi-empirical models based on molecular mechanics have the advantages of easier rational interpretation of binding modes, and they are more sensitive to protein conformational changes. Frequently used energetic terms include van der Waals energy, electrostatic energy, hydrogen-bond energy, desolvation energy, hydrophobic interaction, torsional entropy, and so on [4, 5]. Among these terms, the atomic partial charges of biomolecules are considered of central importance, because they are essential for evaluation of the long-ranged electrostatic interaction, which is known to be a key factor for biomolecular association. Due to the extremely low computational cost, current molecular docking programs often use regression models with distance-dependent molecular descriptors or energy terms to predict the possible binding poses and to evaluate the binding affinity of a small molecule. Such descriptors are also used for large-scale virtual chemical library screening to rapidly narrowing down the chemical space and for subsequent identification of potential drugs.

The affinity of most single glycan–protein interactions is generally low, with K_d values of mM to μM levels [1]. In nature, many GBPs are oligomeric or membrane-associated proteins, which allow aggregation of the GBP in the plane of the membrane. Many of the glycan ligands for GBPs are also multivalent. The interaction of multiple subunits with a multivalent display of glycans raises the affinity of the interaction by several orders of magnitude under the physiological conditions. However, most of the currently used scoring functions may not have comparable performance for the individual "weak binder" as for small molecules with submicromolar to picomolar affinities [6]. There is a thirst for a general scoring function that has equivalent performance on the weak interactions between glycans and GBP.

Robust scoring functions for protein–ligand interactions with quantum chemical charge models

In a previous study, we have employed two well-established quantum chemical approaches, namely the restrained electrostatic potential (RESP) and the Austin-model 1-bond charge correction (AM1-BCC) methods, to obtain atomic partial charges [7] for deriving new scoring functions for the automated molecular docking software package, AutoDock4 [8], which has been widely adopted in virtual screening of drug candidates and prediction of ligand binding poses in protein pockets.

The AutoDock4 scoring function comprises five energetic terms: the van der Waals interactions, the hydrogen bonding interactions, the electrostatic interactions, the desolvation energy, and the torsional entropy. The AutoDock4 scoring function predicts the binding free energy with the following formula:

$$\begin{aligned}\Delta G_{bind} = {} & W_{vdw} \times \sum_{i,j}\left(\frac{A_{ij}}{r_{ij}^{12}} - \frac{B_{ij}}{r_{ij}^{6}}\right) \\ & + W_{H-bond} \times \sum_{i,j} E(t)\left(\frac{C_{ij}}{r_{ij}^{12}} - \frac{D_{ij}}{r_{ij}^{10}}\right) \\ & + W_{estat} \times \sum_{i,j} \frac{q_i q_j}{\varepsilon(r_{ij}) r_{ij}} \\ & + W_{desol} \times \sum_{i,j} (S_i V_j + S_j V_i) e^{\left(-r_{ij}^2/2\sigma^2\right)} \\ & + W_{tor} \times N_{tors}\end{aligned}$$

The atomic charges used to evaluate the electrostatics energy term of the original 2007 AutoDock4 scoring function were calculated using the Gasteiger charge model [9], whose primary advantages lie in its simplicity and speed. However, such charge calculations can generate atomic charges that are less accurate than those determined by quantum chemical methods.

We implemented two variants of AutoDock4 scoring functions using two well-established charge models for ligands, namely, RESP [9] and AM1-BCC [11 – 12], that have been used widely in molecular dynamics simulations with the AMBER force field. RESP is a two-stage restrained electrostatic fit charge model, while AM1-BCC is a quick and efficient semi-empirical atomic charge model that aims to achieve the accuracy of RESP. The atomic charges of proteins were retrieved from the AMBER parm99SB force field parameters, which were mainly derived by the RESP methodology [13 – 15]. The abbreviations "AP" for AM1-BCC (ligand)/Amber PARM99SB (protein) and "RP" for RESP (ligand)/Amber PARM99SB (protein) will be used in the following sections.

In combination with robust regression analysis and outlier exclusion, our protein–ligand free energy regression with the robust AP (RAP) charge model achieves lowest root-mean-squared error of 1.637 kcal/mol for the training set of 147 complexes and 2.176 kcal/mol for the external test set of 1.427 complexes. The assessment for binding pose prediction with the 100 external decoy sets indicates very high success rate of 87% with the criteria of predicted root-mean-squared deviation of less than 2 Å (Table 1 and Figure 1). The success rates and statistical performance of our robust scoring functions are only weakly dependent on the type of protein–ligand interactions (Table 2).

Table 1. Success rates of binding site prediction by different scoring functions[a] [7]

	success rate (%) for different RMSD criteria				
scoring function	≦1Å	≦1.5Å	≦2Å	≦2.5Å	≦3Å
DrugScore^{CSD}	83	85	87		
AutoDock4^{RAP}	83	85	87	87	87
AutoDock4^{RGG}	80	82	86	86	86
AutoDock4^{RRP}	79	81	84	85	85
original AutoDock4^{GG}	74	76	79	79	79
Cerius2/PLP	63	69	76	79	80
SYBYL/F-Score	56	66	74	77	77
Cerius2/LigScore	64	68	74	75	76
DrugScore	63	68	72	74	74
Cerius2/LUDI	43	55	67	67	67
X-Score	37	54	66	72	74
AutoDock3	34	52	62	68	72
Cerius2/PMF	40	46	52	54	57
SYBYL/G-Score	24	32	42	49	56
SYBYL/ChemScore	12	26	35	37	40
SYBYL/D-Score	8	16	26	30	41

[a] Except for the results of the AutoDock4 scoring functions, the results of DrugScore^{CSD} and other scoring functions were taken from Velec *et al.*[26][18] and Wang *et al.* [19], respectively.
[b] Scoring functions are sorted by the number of cases under 2Å.

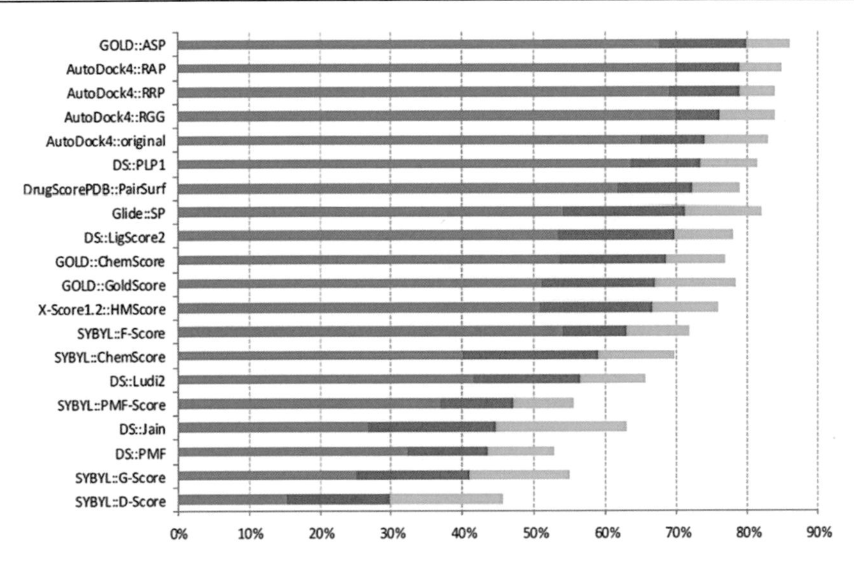

Figure 1. Comparison of the success rates of AutoDock4 scoring functions and 16 scoring functions provided by Cheng *et al* [20]. The cutoffs are rmsd < 1.0 Å (blue bars), < 2.0 Å (red bars), and < 3.0 Å (green bars), respectively. The native binding poses of ligands were included in the decoy sets. Scoring functions are sorted by the number of cases under 2 Å [7].

Table 2. Success rates of binding pose prediction of various scoring functions[a] on three classes of complexes [7]

	success rate (%; RMSD ≦2Å)			
	Overall	hydrophilic	mixed	hydrophobic
scoring function	(100)	(44)	(32)	(24)
AutoDock4^{RAP}	87	89	91	79
AutoDock4^{RGG}	86	86	91	79
AutoDock4^{RRP}	84	84	91	75
original AutoDock4^{GG}	79	77	81	79
Cerius2/PLP	76	77	78	71
SYBYL/F-Score	74	75	75	71
Cerius2/LigScore	74	77	75	67
DrugScore^{PDB}	72	73	81	58
Cerius2/LUDI	67	75	66	54
X-Score	66	82	59	46
AutoDock3	62	73	53	54
Cerius2/PMF	52	68	44	33
SYBYL/G-Score	42	55	34	29
SYBYL/ChemScore	35	32	34	42
SYBYL/D-Score	26	23	28	29

[a] Data were adopted from Wang *et al.*[19] except for AutoDock4 scoring functions.
[b] Scoring functions are sorted according to the overall success rates.

Recognition of glycan by proteins is a key to the specificity in glycobiology

Binding of glycans to proteins represents the major way in which the information contained in glycan structures is recognised, deciphered, and put into biological action [1]. The structures of hundreds of glycan–protein complexes have been determined by X-ray crystallography and NMR spectroscopy. In most cases, the glycan-binding sites typically accommodate one to four sugar residues. Unveiling the three-dimensional structure of a glycan–protein complex can reveal much about the specificity of binding, changes in conformation that take place on binding, and the contribution of specific amino acids to the interaction.

Hydrophobic interactions are very common in glycan–protein complexes and can involve aromatic residues as well as alkyl side chains of amino acids in the binding pocket [1]. Since the forces involved in the binding of a glycan to a protein are the same as for the binding of a ligand to its receptor (hydrogen bonding, electrostatic or charge interactions, van der Waals interactions, and dipole attraction), it is tempting to try to calculate their contribution to overall binding energy. Unfortunately, calculating the free energy of association is difficult for several reasons, including problems in defining the conformation of the unbound versus the bound glycan, changes in bound water within the glycan and the binding site, and conformational changes in the GBP upon binding. To take the first step to tackle these problems, we tested the capability of our established AutoDock4^{RAP} scoring function to predict the binding affinities of glycans to GBP.

Performance of AutoDock4^{RAP} on predicting binding affinities of glycans to GBP

GBP can be broadly classified into two major groups: glycosaminoglycan-binding proteins and lectins. Because glycosaminoglycan-binding proteins do not have shared structural features, we applied the AutoDock4^{RAP} scoring function to the crystal structures of glycan–lectin complexes for which the binding affinities have been determined experimentally [16].

Lectins tend to recognise specific terminal aspects of glycan chains by fitting them into shallow but relatively well-defined binding pockets, namely, "carbohydrate-recognition domains" (CRD) that often retain specific features of primary amino acid sequence or three-dimensional structure [1]. The binding affinities to a single CRD in many lectins appear to be low (with K_d values in the micromolar range).

During the initial preparation work on the 23 complex structures for subsequent docking, we did not include the crystal structure with PDB code 1EN2 because the frequently appeared missing residues in the protein coordinates led to abrupt termination of the process. Among the 22 crystal structures used in the current validation study (Table 3), four complexes have glycosylated residues (1AXO, 1AX1, 1AX2, and 1AXZ). The covalently linked oligosaccharides are excluded from the analyses since they do not serve as ligands for the proteins.

Table 3. Validation of AutoDock4^{RAP} on glycan–lectin complexes.
The AutoDock4^{RAP} scoring function was applied to the crystal structures of glycan–lectin complexes for which the binding affinities have been determined experimentally [15]. The refined ligand binding modes of the complexes used in the study have root-mean-square deviations (RMSD) no more than 1.21 Å in reference to the corresponding crystal binding modes, and the refined free energy of binding for the 22 glycan–lectin complexes has a root-mean-squared error of 1.606 kcal/mol in reference to the experimental values. [a] Complex with crystal packing effect at the binding site.

PDB ID	Protein name	ΔG_{exp}	Rescore	Refine		Docking ^{rank1}		Docking ^{rank2}		Docking ^{rank3}		Ligand in crystal
		(kcal/mol)	ΔG	ΔG	RMSD	ΔG	RMSD	ΔG	RMSD	ΔG	RMSD	
1J4U[a]	Artocarpin	-4.36	-2.95	-3.69	0.73	-7.41	26.95	-6.65	1.35	-6.48	26.76	O1-Methyl-Mannose
5CNA[a]	Concanavalin A	-5.31	-4.35	-4.76	0.35	-6.53	0.98	-5.84	16.85	n.a.	n.a.	O1-Methyl-Mannose
1GIC[a]	Concanavalin A	-4.61	-4.57	-4.82	0.36	-6.37	0.77	-6.13	1.75	-5.46	17.51	Methyl-α-D-Glucopyranoside
1QDO[a]	Concanavalin A	-6.81	-3.35	-4.70	0.69	-7.26	3.12	-6.92	2.05	-6.59	2.94	(α-D-Man)-(O1-Methyl-Man)
1QDC[a]	Concanavalin A	-5.31	-4.20	-4.52	0.51	-8.18	16.57	-6.55	2.48	-6.26	1.24	(α-D-Mannose)-(O1-Methyl-Mannose)
1ONA[a]	Concanavalin A	-7.41	-1.54	-5.50	0.50	-6.83	2.32	-6.82	1.74	-5.57	3.75	(α-D-Mannose)-(O1-Methyl-Mannose)-(α-D-Mannose)
1DGL[a]	Lectin	-8.21	-4.41	-5.24	0.37	-6.77	1.96	-6.47	5.25	-6.26	2.36	(α-D-Mannose)-(O1-Methyl-Mannose)-(α-D-Mannose)
1AXZ[a]	Lectin	-4.35	-2.33	-3.46	0.81	-6.27	17.32	-5.54	0.99	-5.18	10.11	α-D-Galactose; β-D-Galactose
1AX0[a]	Lectin	-4.28	-3.36	-4.86	0.62	-7.22	16.10	-6.72	1.11	-5.93	16.49	N-Acetyl-2-Deoxy-2-Amino-Galactose
1AX1[a]	Lectin	-4.50	-1.75	-2.43	0.69	-7.32	11.97	-5.86	3.11	-5.08	2.66	(β-D-Glucose)-(β-D-Galactose)
1AX2[a]	Lectin	-5.43	-1.90	-2.50	0.90	-5.97	18.72	-5.93	20.84	-5.38	3.26	[2-(Acetylamino)-2-Deoxy-A-D-Glucopyranose]-[β-D-Galactose]
2BQP	Lectin	-3.35	-3.09	-3.84	0.42	-6.35	0.91	-5.69	19.18	-5.19	19.39	α-D-Glucose
1BQP	Lectin	-3.97	-3.74	-4.14	0.28	-7.23	8.30	-6.50	19.04	-6.47	9.76	$(\alpha\text{-D-Mannose})_2$
1QF3	Agglutinin	-4.06	-1.53	-2.85	0.79	-6.77	11.82	-5.14	9.12	-5.00	1.88	Methyl-β-Galactose
2PEL[a]	Agglutinin	-4.25	-1.16	-2.66	0.63	-6.51	2.76	-5.05	3.04	-4.65	9.12	α-Lactose (LBT) *3 +β-Lactose (LAT) *1
1EHH[a]	Agglutinin isolectin VI	-5.11	-0.97	-2.81	1.21	-6.34	13.14	-5.09	12.71	-4.54	15.03	$(\text{N-Acetyl-D-Glucosamine})_3$
1K7U	Agglutinin isolectin III	-5.11	-0.28	-3.58	1.01	-6.59	16.61	-5.35	17.34	-5.33	13.93	$(\text{N-Acetyl-D-Glucosamine})_2$
1KUJ[a]	Agglutinin	-4.14	-4.27	-5.40	0.45	-6.95	11.14	-6.07	10.41	-5.83	12.84	O1-Methyl-Mannose
1GZC	Lectin	-4.76	-1.60	-2.42	0.73	-7.54	11.53	-6.45	18.26	-5.81	19.97	β-Lactose
1HKD[a]	Lectin	-3.83	-3.45	-3.89	0.25	-6.21	19.16	-5.03	7.26	-5.02	1.14	Methyl-α-D-Glucopyranoside
4GAL[a]	Human galectin-7	-4.62	+16.23	-2.95	1.06	-7.06	11.17	-5.86	10.9	-4.65	9.55	(β-D-Galactose)-(β-D-Glucose)
5GAL[a]	Human galectin-7	-4.40	+1.23	-2.76	0.95	-6.60	12.88	-6.46	9.52	-6.32	7.58	(N-Acetyl-D-Glucosamine)-(β-D-Galactose)

In the analyses of free energy of binding, we carried out three stages of measures. The first one was to "rescore" the original binding mode in the crystal coordinates without moving the ligand. Next, we allowed the ligand to move in a restricted space using local search parameters, without torsional or rotational modification; thereby "refining" the ligand to a potential position with lower binding free energy in the crystal binding site. The refined ligand binding modes of the complexes used in the study have root-mean-square deviations

(RMSD) no more than 1.21 Å in reference to the corresponding crystal binding modes (Table 3). The refined free energy of binding for the 22 glycan–lectin complexes has a root-mean-squared error of 1.606 kcal/mol in reference to the experimental values.

We also carried out a comprehensive search, rendering the ligand to have translational and rotational alterations, "docking" the ligand to a larger space in the binding pocket of the protein. Because the lectin structures used in the study all have shallow CRD with relatively large area, we enclosed the entire CRD for the docking analysis of each complex.

As we inspected the structures of these glycan–lectin complexes, we found out that 17 out of the 22 complexes used in the study have crystal packing effects at the binding sites, that is, the ligand (glycan) bound to the protein (lectin) at the interface of different symmetry mates when we generated them using the crystallographic symmetry information (Table 3 and Figure 2). The crystal packing effect could be indicative of an artefact in the crystal binding mode for certain complex structures. For the glycan–lectin complexes, however, the crystal packing effects at the binding sites do not necessarily reflect the dockability of the glycan ligands. Using the AutoDock4^{RAP} scoring function, we can still reproduce the crystal binding modes (with RMSD less than 2 Å) after comprehensive docking analyses on many of the complexes with such crystal packing effect (Table 3 and Figures 3 – 5). This could be due to the spatial arrangement in the recognition of glycan by the shallow CRD of lectin, which is quite different from that of the proteins with deep ligand binding pockets.

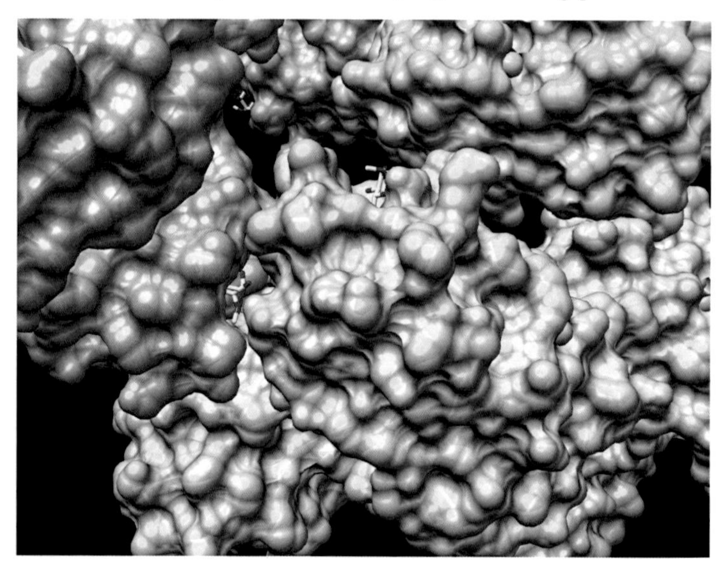

Figure 2. Representative image of crystal packing effects at the binding site. The symmetry mates of artocarpin (PDB code: 1J4U) were generated using the crystal symmetry information and are shown in different colours. The O1-methyl-mannose bound to the interface of the violet and grey molecules indicates crystal packing effect at this ligand-binding site.

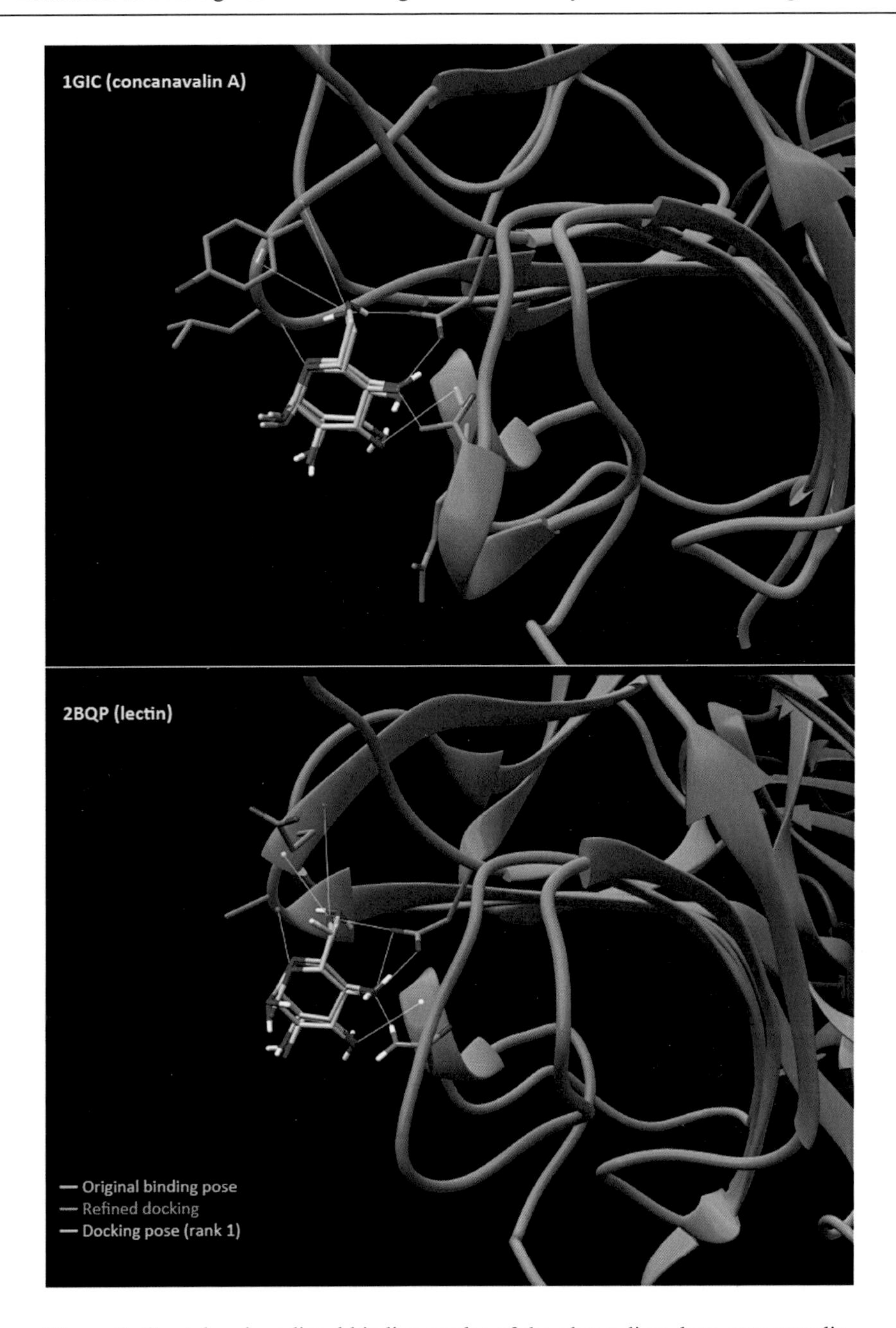

Figure 3. Crystal and predicted binding modes of the glycan ligands to concanavalin A (1GIC) and lectin (2BQP). Cyan lines indicate the hydrogen bonds formed between the glycan ligand and the protein.

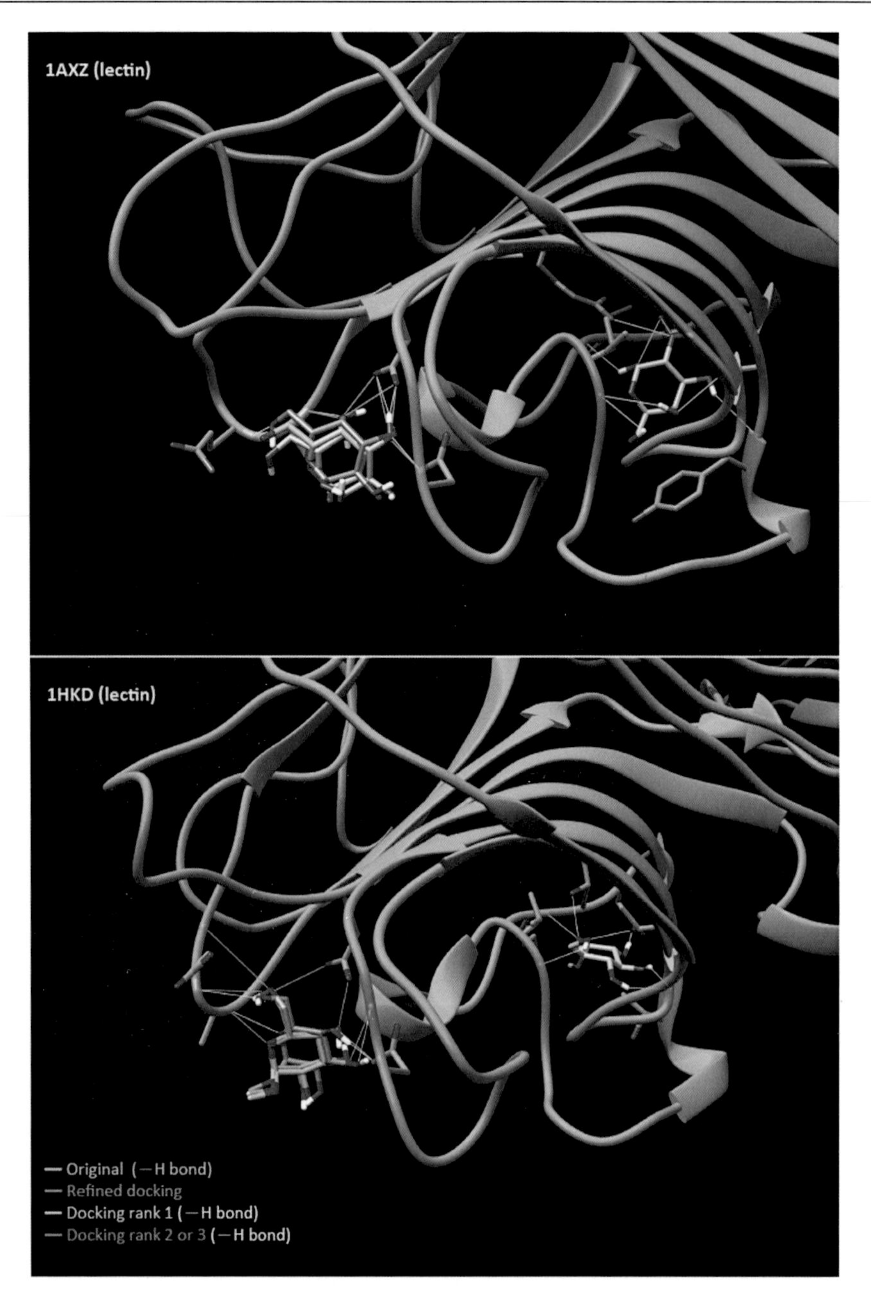

Figure 4. Crystal and predicted binding modes of the glycan ligands to lectins (1AXZ and 1HKD). Cyan lines indicate the hydrogen bonds formed between the protein and the ligand in the crystal binding mode, while yellow lines indicate the hydrogen bonds formed between the ligand in the predicted binding modes.

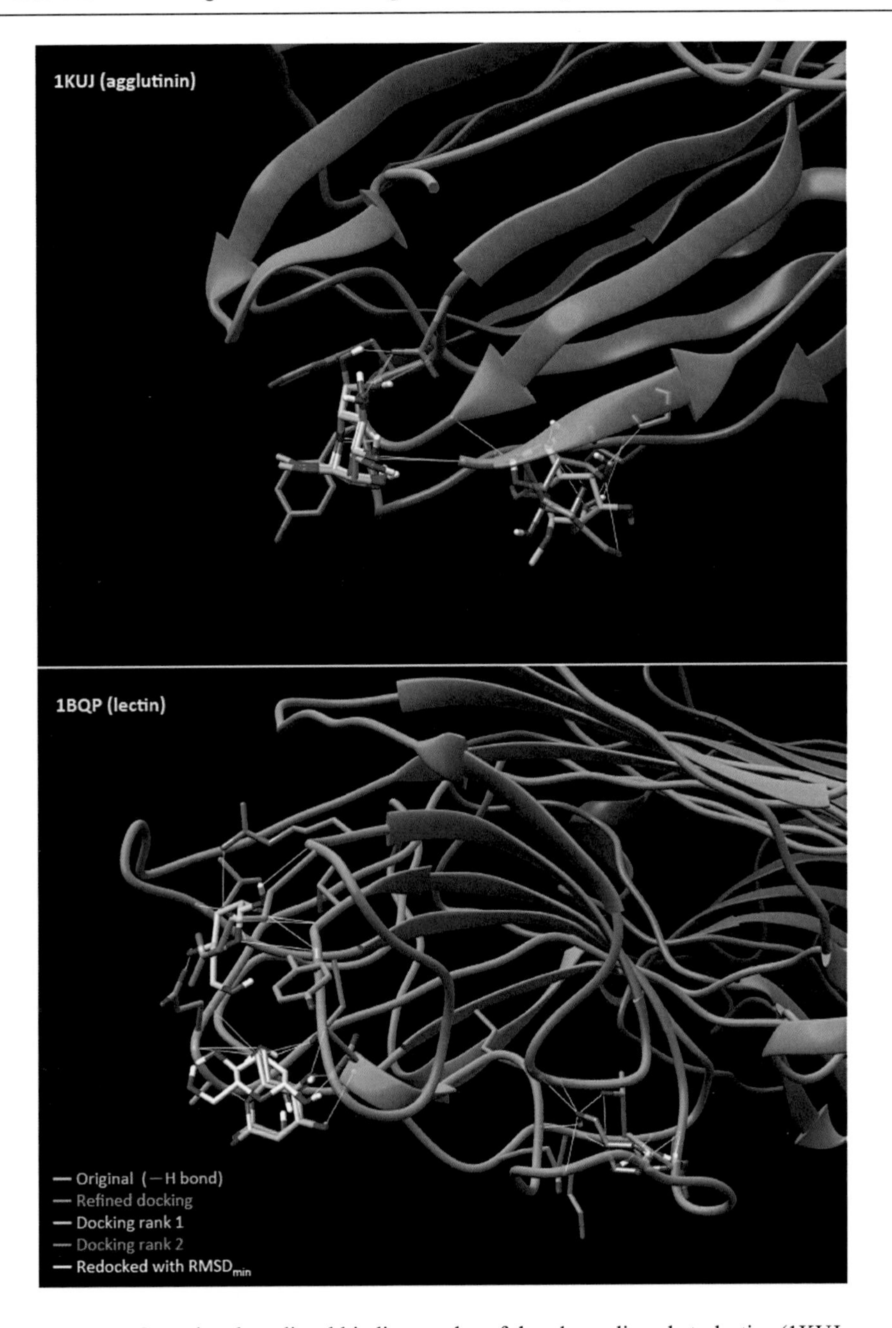

Figure 5. Crystal and predicted binding modes of the glycan ligands to lectins (1KUJ and 1BQP). Cyan lines indicate the hydrogen bonds formed between the protein and the ligand in the crystal binding mode, while yellow lines indicate the hydrogen bonds formed between the ligand in the predicted binding modes.

Potential application of AutoDock4^{RAP} on the glycan–GBP systems

During the initial development of the molecular biology, studies of glycans lagged far behind those of other major classes of molecules [1]. This was in large part due to their inherent structural complexity, the great difficulty in determining their sequences, and the fact that their biosynthesis could not be directly predicted from a DNA template. This delayed development in experimental methodology could reflect the performance of computational work on glycobiology. Kerzmann *et al.* presented a method specifically designed for the docking of carbohydrate-like compounds [15]. In contrast, although the AutoDock4^{RAP} scoring function was not tailored for glycan–GBP complexes, the current validation study revealed the capability of AutoDock4^{RAP} to predict the binding affinities for GBP. The use of a general, robust scoring function can facilitate the virtual screening on compounds with more diverse chemical scaffolds, the essential work in the very beginning of rational drug design, for the various GBP.

One of the critical issues in calculating the free energy of binding lies in the conformational changes in the protein (GBP) upon ligand (glycan) binding. This can be addressed using the relaxed complex scheme [17 – 19] since the atomic charge models used in the current scoring function are those have been widely used in molecular dynamics simulations with the AMBER force field.

In the complexes used in the study, there are a considerable number of hydrogen bonds between the glycans and lectins, as demonstrated in both the crystal and predicted binding modes (Figures 3 – 5). The contribution of hydrogen bonding to the binding affinities of glycans to GBP can be further examined with energy decomposition analysis when the relaxed complex scheme is used. Nevertheless, the amino acid residues with potentials to form hydrogen bonds with glycans may also serve as target residues in the design of lead compounds to inhibit glycan–GBP interactions.

SUMMARY

The physiological and pathological significances of glycan–binding proteins are drawing more and more attention, both in basic and applied sciences. In the current study, we have demonstrated the capability of a general, robust scoring function, AutoDock4^{RAP}, to predict the binding affinities for glycan–binding proteins, without any calibration to this specific class of protein-ligand interactions. The free energy of binding for the 22 glycan–lectin complexes has a root-mean-squared error of 1.606 kcal/mol in reference to the experimental values. The use of AutoDock4^{RAP} can therefore facilitate the virtual screening on compounds with more diverse chemical scaffolds, as well as further rigorous studies, such as those with use of relaxed complex scheme and energy decomposition analysis.

References

[1] Varix A., Cummings R.D., Esko J.D., Freeze, H.H., Stanley, P., Bertozzi, C. R., Hart, G.W., Etzler, M.E., Eds. (2009) *Essentials of Glycobiology*. 2nd edition. Cold Spring Harbor Laboratory Press: Cold Spring Harbor (NY).

[2] Gilson, M.K., Zhou, H.-X. (2007) Calculation of Protein–Ligand Binding Affinities. *Ann. Rev. Biophys. Biomol. Struc.* **36**:21 – 42. doi: 10.1146/annurev.biophys.36.040306.132550.

[3] Gilson, M.K., Given, J.A., Bush, B.L., McCammon, J.A. (1997) The statistical–thermodynamic basis for computation of binding affinities: A critical review. *Biophysical J.* **72**:1047 – 1069. doi: 10.1016/S0006-3495(97)78756-3.

[4] Morris, G.M., Goodsell, D.S., Halliday, R.S., Huey, R., Hart, W.E., Belew, R.K., Olson, A.J. (1998) Automated docking using a Lamarckian genetic algorithm and an empirical binding free energy function. *J. Comp. Chem.* **19**:1639 – 1662. doi: 10.1002/(sici)1096-987x(19981115)19:14<1639::aid-jcc10>3.0.co;2-b.

[5] Huey, R., Morris, G.M., Olson, A.J., Goodsell, D.S. (2007) A semi-empirical free energy force field with charge-based desolvation. *J. Comp. Chem.* **28**:1145 – 1152. doi: 10.1002/jcc.20634.

[6] Wang, J.C., Lin, J.H. (2013) Scoring functions for prediction of protein-ligand interactions. *Curr. Pharm. Des.* **19**:2174 – 2182. doi: 10.2174/1381612811319120005.

[7] Wang, J.-C., Lin, J.-H., Chen, C.-M., Perryman, A.L. Olson, A.J. (2011) Robust scoring functions for protein-ligand interactions with quantum chemical charge models. *J. Chem. Inf. Model.* **51**:2528 – 2537. doi: 10.1021/ci200220v.

[8] Morris, G.M., Huey, R., Lindstrom, W., Sanner, M.F., Belew, R.K., Goodsell, D.S., and Olson, A.J. (2009) AutoDock4 and AutoDockTools4: Automated Docking with Selective Receptor Flexibility. *J. Comput. Chem.* **30**(16):2765 – 2791. doi: 10.1002/jcc.21256.

[9] Gasteiger, J., Marsili, M. (1980) Iterative partial equalization of orbital electronegativity – a rapid access to atomic charges. *Tetrahedron* **36**:3219 – 3228. doi: 10.1016/0040-4020(80)80168-2.

[10] Bayly, C.I., Cieplak, P., Cornell, W., Kollman, P.A. (1993) A well-behaved electrostatic potential based method using charge restraints for deriving atomic charges – the resp model. *J. Phys. Chem.* **97**:10269 – 10280. doi: 10.1021/j100142a004.

[11] Jakalian, A., Bush, B.L., Jack, D.B., Bayly, C.I. (2000) Fast, efficient generation of high-quality atomic charges. AM1-BCC model: I. Method. *J. Comp. Chem.* **21**:132 – 146.
doi: 10.1002/(SICI)1096-987X(20000130)21:2<132::AID-JCC5>3.0.CO;2-P

[12] Jakalian, A., Jack, D.B., Bayly, C.I. (2002) Fast, efficient generation of high-quality atomic charges. AM1-BCC model: II. Parameterization and validation. *J. Comp. Chem.* **23**:1623 – 1641.
doi: 10.1002/jcc.10128.

[13] Cornell, W.D., Cieplak, P., Bayly, C.I., Gould, I.R., Merz, K.M., Ferguson, D.M., Spellmeyer, D.C., Fox, T., Caldwell, J.W., Kollman, P.A. (1995) A second generation force-field for the simulation of proteins, nucleic acids, and organic molecules. *J. Am. Chem. Soc.* **117**:5179 – 5197.
doi: 10.1021/ja00124a002.

[14] Ponder, J.W., Case, D.A. (2003) Force fields for protein simulations. *Adv. Prot. Chem.* **66**:27 – 85.
doi: 10.1016/S0065-3233(03)66002-X.

[15] Duan, Y., Wu, C., Chowdhury, S., Lee, M.C., Xiong, G., Zhang, W., Yang, R., Cieplak, P., Luo, R., Lee, T., Caldwell, J., Wang, J.M., Kollman, P.A. (2003) A point-charge force field for molecular mechanics simulations of proteins based on condensed-phase quantum mechanical calculations. *J. Comp. Chem.* **24**:1999 – 2012.
doi: 10.1002/jcc.10349.

[16] Kerzmann, A., Fuhrmann, J., Kohlbacher, O., Neumann, D. (2008) BALLDock/SLICK: A new method for protein-carbohydrate docking. *J. Chem. Inf. Model.* **48**:1616 – 1625.
doi: 10.1021/ci800103u.

[17] Lin J.H., Perryman A.L., Schames J.R., McCammon J.A. (2002) Computational drug design accommodation receptor flexibility: The relaxed complex method. *J. Am. Chem. Soc.* **68**:47 – 62.

[18] Lin J.H., Perryman A.L., Schames J.R., McCammon J.A. (2003) The relaxed complex method: Accommodating receptor flexibility for drug design with an improved scoring scheme. *Biopolymers* **68**:47 – 62.
doi: 10.1021/ja0260162.

[19] Lin J.H. (2011) Accommodating protein flexibility for structure-based drug design. *Curr. Top. Med. Chem.* **11**:171 – 178.
doi: 10.2174/156802611794863580

[20] Cheng, T., Li, X., Li, Y., Liu, Z., Wang, R. (2009). Comparative Assessment of Scoring Functions on a Diverse Test Set. *J. Chem. Inf. Model.* **49**:1079 – 1093. doi: 10.1021/ci9000053.

Glycomics and Glycoproteomics Databases in Japan and Asia

Hisashi Narimatsu

Research Center for Medical Glycoscience (RCMG), National Institute of Advanced Industrial Science and Technology (AIST), Central 2, 1 – 1-1, Umezono, Tsukuba, Ibaraki, 305 – 8568, Japan.

E-Mail: h.narimatsu@aist.go.jp

Received: 22nd January 2014 / Published: 22nd December 2014

Abstract

The "Integrated Database Project" was initiated to establish a publically accessible database to integrate all useful life science databases in Japan. Our JCGGDB (Japan Consortium for Glycobiology and Glycotechnology Database) was selected as a promotion program in the project, focussing on the integration of all the glycan-related databases and establishment of user-friendly search systems. As part of the project, we intend the integration of databases not only within Asia but also with other countries. Working closely with various institutes in Japan and the world, we continuously develop base technologies for the database integration, facilitate interactions between databases in the field of glycoscience as well as other associated study areas, and build bioinformatics tools to support experimental study. Our goal is to create a truly useful database that could be easily and intuitively understood by every user.

Introduction

Since 2001, several glycoscience projects have been initiated by the New Energy and Industrial Technology Development Organization (NEDO) in Japan. AIST RCMG has been playing the central role in these projects. The first project was named the "Glycogene Project (GG-P)" and was undertaken to comprehensively identify the human glycosyltransferase genes and analyse their substrate specificities. The second project called the

This article is part of the Proceedings of the Beilstein Glyco-Bioinformatics Symposium 2013.
www.proceedings.beilstein-symposia.org

"Structural Glycomics Project (SG-P)" was conducted to develop technologies to analyse the structures of glycans. On the basis of the knowledge acquired and the technologies developed via GG-P and SG-P, we successfully conducted the subsequent "Medical Glycomics Project (MG-P)" focusing on the functional analysis and clinical applications of glycans.

In GG-P, we used the latest bioinformatics techniques and comprehensively searched for the glycosyltransferase genes in the database for human genome sequencing, which was almost completed at that time. The obtained data were registered in the in-house database to be used for our ongoing research work.

In 2007, the Ministry of Education, Culture, Sports, Science and Technology (MEXT) of Japan initiated the "Integrated Database Project". This project was designed to establish a publically accessible database that would integrate all the diverse life science databases in Japan. Our glycoscience databases were also included in the project. We started publicising some of our databases and collaborated with other researchers from the field of glycoscience in Japan. At the same time, we changed the name of our database to Japan Consortium for Glycobiology and Glycotechnology Database (JCGGDB). The institutes that collaborated to form JCBBDB are: AIST, Noguchi Institute, Ritsumeikan University, Soka University, and Riken. This project was taken over from MEXT by the Japan Science and Technology Agency (JST). Although the grant from JST for this project will expire in March 2014, a subsequent project will be conducted thereafter (Figure 1).

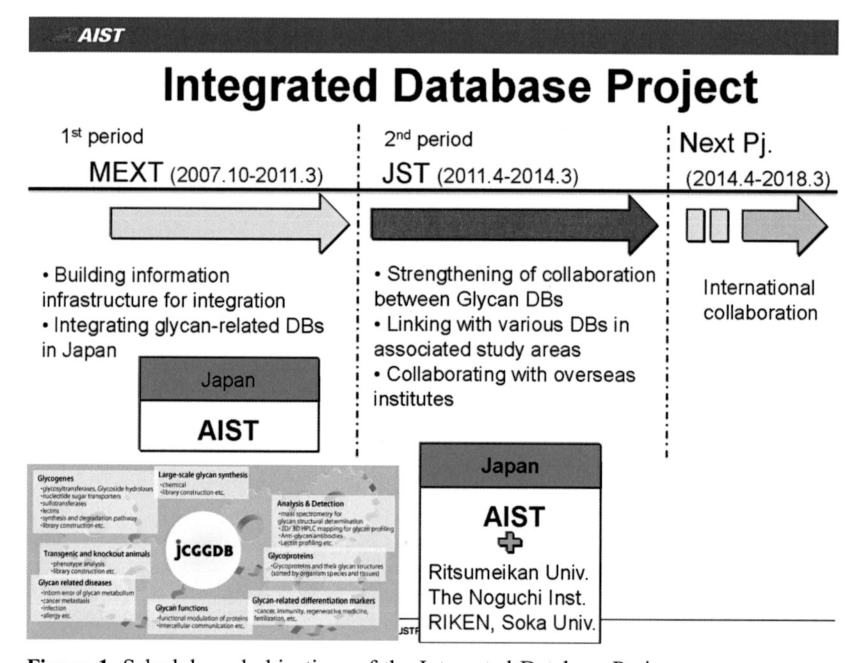

Figure 1. Schedule and objectives of the Integrated Database Project.

Results

Figure 2 shows all the databases we have developed so far. Of these databases, the Glycogene Database (GGDB), Lectin Frontier Database (LfDB), Glycan Mass Spectral Database (GMDB), and Glycoprotein Database (GlycoProtDB) contain the experimental data generated by AIST (Table 1).

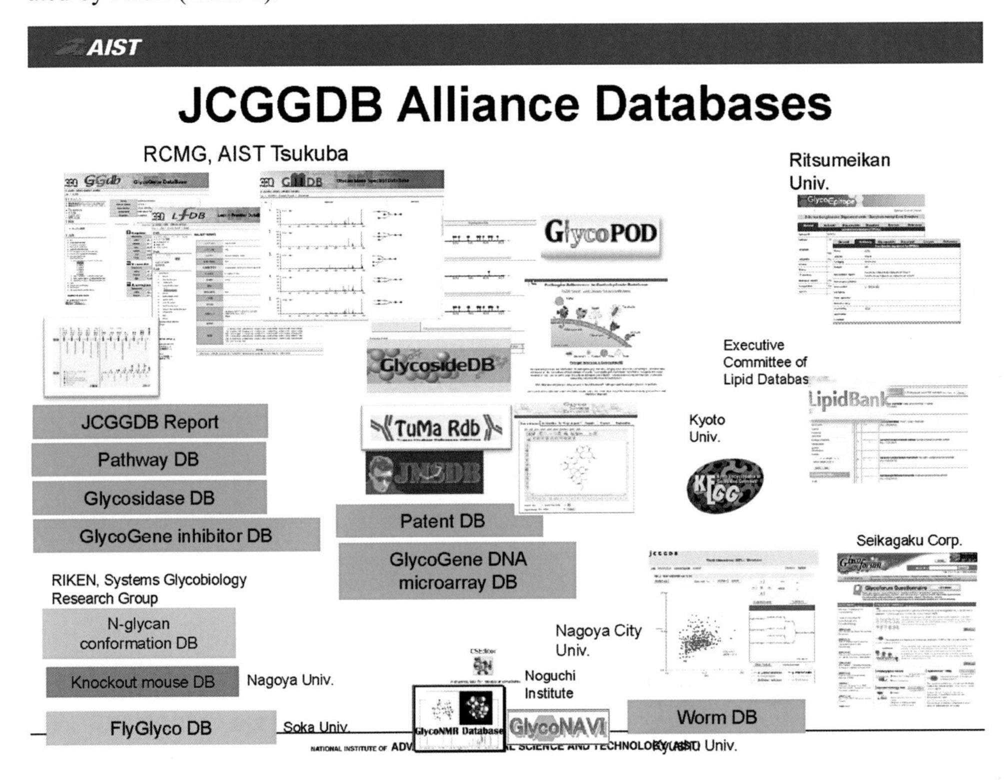

Figure 2. Current JCGGDB Alliance Databases.

Table 1. JCGGDB Database Alliance in Japan.

AIST Databases	JCGGDB Databases
GlycoProtDB	GlycoPOD
LfDB	GlycoEpitope
GMDB	GlycoNAVI/GlycoNMR
GGDB	Glycan Annotation DBs

1. GGDB (with almost 200 entries, Figure 3) provides information on the substrate specificity of glycosyltransferases, sugar-nucleotide synthases, and sugar-nucleotide transporters via the gene-based entries.

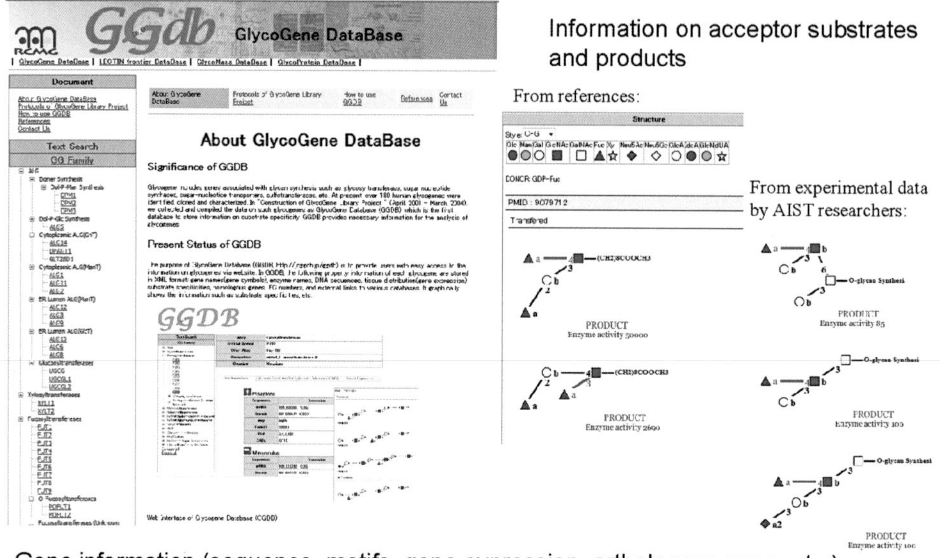

Figure 3. GGDB: GlycoGene Database.

2. GGDB is an outcome of our first NEDO project, GG-P, which originally contained only the known glycosyltransferase genes and their homologs sequences obtained through the initial human genome sequencing for in-house use during daily research. Information on other genes was subsequently collected from external databases and research papers to establish a publically accessible database. Many similar databases with information on genes are currently available on the web. Therefore, we are trying to extensively revise this database by focusing on the enzymatic and biological relevance of the glycogenes.
3. LfDB (with almost 80 entries, Figure 4) provides affinity constants of lectins and glycans measured using frontal affinity chromatography. More than 80 lectins immobilised onto the columns were tested using more than 100 referential glycan compounds. The number of the entries for the affinity data is continuously growing.

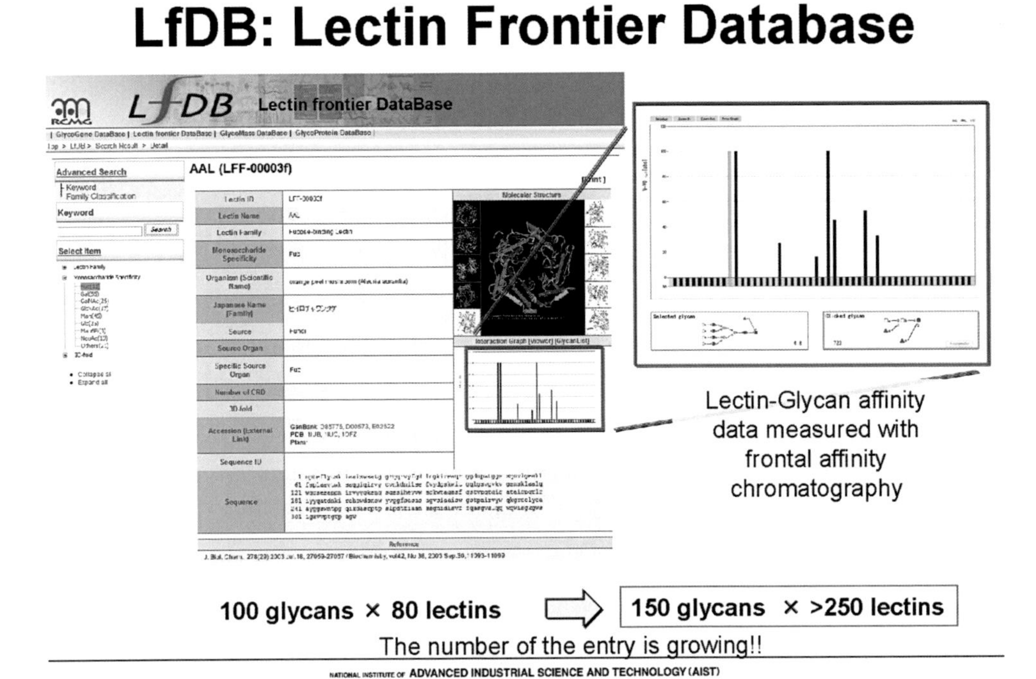

Figure 4. LfDB: Lectin Frontier Database.

4. GMDB (with almost 3000 entries, Figure 5) provides comparable mass spectrometry (MS) profiles of isomeric glycan molecules. The data were obtained from as many glycan structures as possible, including those of the commercially acquired referential compounds; in cases where no commercial referential compounds were available, we synthesised the compounds in our lab by enzymatic or organic chemical synthesis. GMDB contains the *m/z* values of tandem-MS obtained by performing matrix-assisted laser desorption ionisation (MALDI)-MS, which enables the structural analysis including identification of isomers.

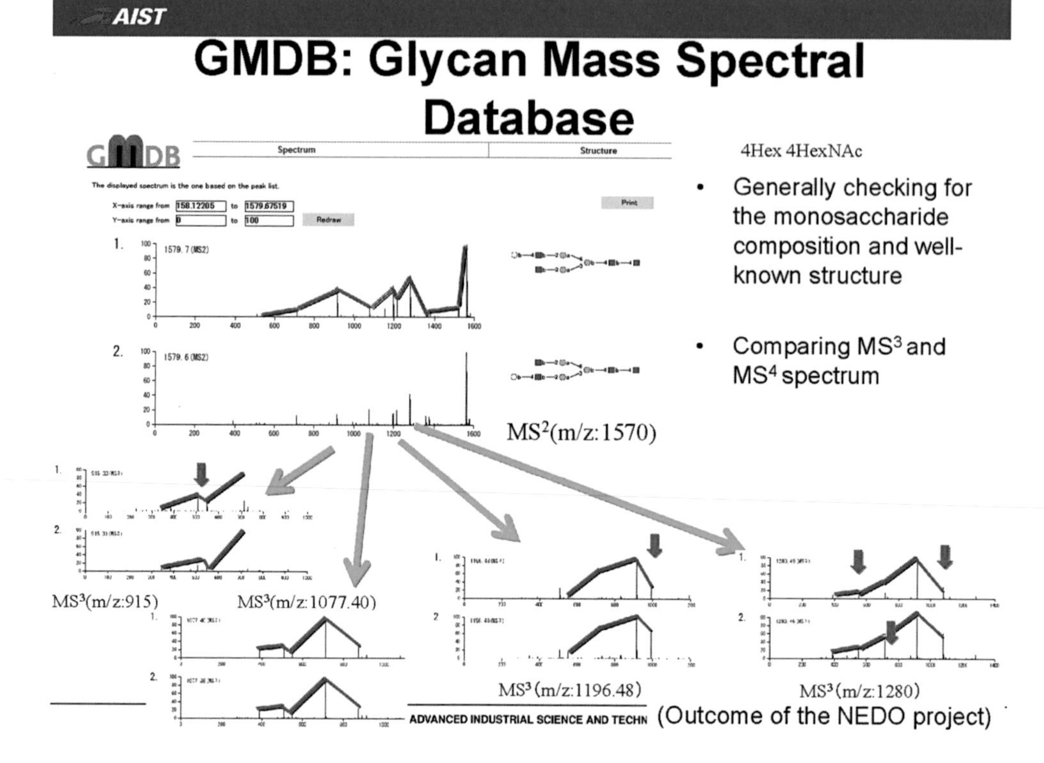

Figure 5. GMDB: Glycan Mass Spectral Database.

5. GlycoProtDB (with almost 3000 entries, Figure 6) contains data on the attachment sites of *N*-glycans obtained experimentally through the proteomic analysis of *N*-glycosylated glycoproteins in humans, mouse, nematode, and drosophila.

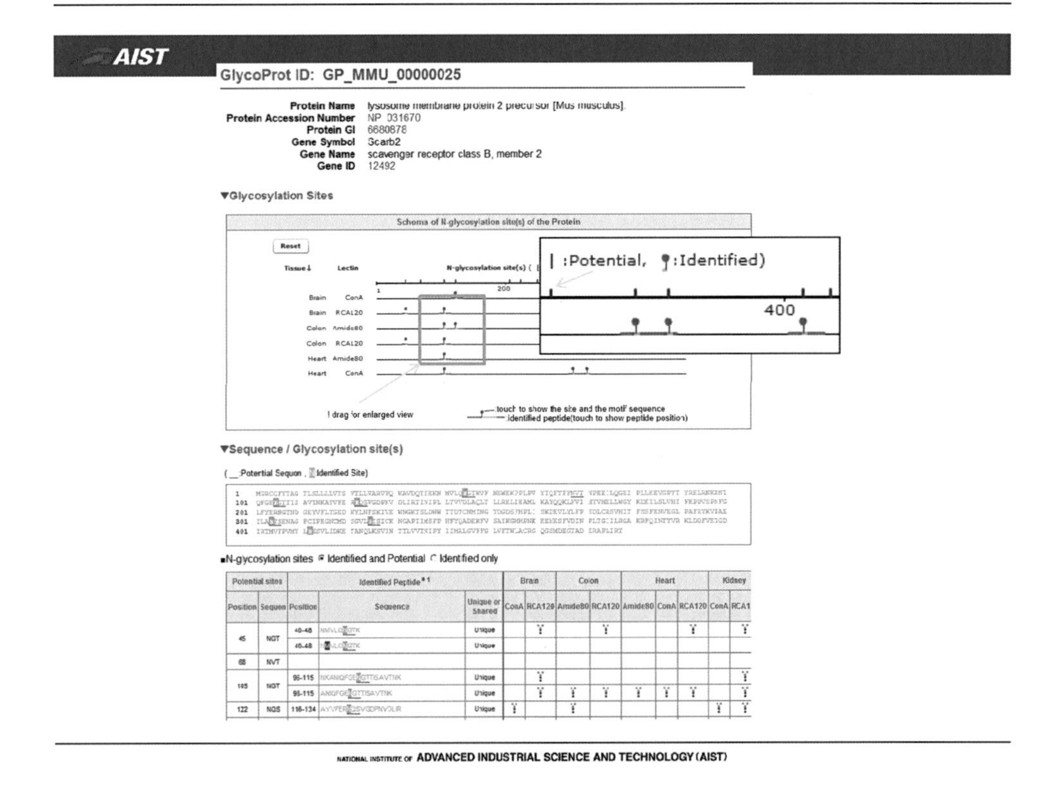

Figure 6. GlycoProt DB: Glycoprotein Database.

The principle technology is based on the LC/MS shotgun protein identification method, but we selectively enrich and concentrate the glycopeptides from crude samples by using affinity columns with lectin probes. Then *N*-linked glycans on the glycopeptides are enzymatically isolated using stable-isotope-labelled water ($H_2{}^{18}O$) and are labelled with the isotope. During this reaction, asparagines isolated from the *N*-linked glycans are converted into aspartic acid containing ^{18}O. This technology, named "lectin-IGOT-LC/MS" (LC/MS spectrometry combined with lectin-mediated affinity capture and isotope-coded glycosylation with site-specific tagging), improves the reliability of the MS analysis. The obtained amino acid sequences of the glycopeptides are mapped with respect to the complete protein sequences; then the matched results are shown by the GlycoProtDB. The pins shown on the amino acid sequences in the figure indicate the active glycosylation sites of *N*-linked glycans identified by the IGOT method. The glycopeptide enrichment required in this method also provides information on the lectin specificity of *N*-glycans, which would be very useful for determining *N*-glycan structures if combined with the data in the LfDB.

The GlycoProtDB also contains metadata on tissues and lectin columns used for sample preparation. Thus, this database also provides the data on glycoproteins (glycoforms) specific to the original tissues.

The databases not derived from the experimental data of AIST

Other than the four databases mentioned above, we also offer access to many other useful databases (Figure 2 and Table 1), e.g.,

1. Glycan Structure Database (with almost 30000 entries) houses the glycan structures collected from the literature.
2. Monosaccharide Database (with almost 800 entries) contains the chemical structural formula of monosaccharides.
3. Pathogen Adherence to Carbohydrate Database (PACDB, with almost 1700 entries) provides information on pathogens (e.g., bacteria, fungi, and viruses and their toxins) adhering to the carbohydrates expressed on the cell surface of host animals or plants.
4. Tumour Markers Reference Database (TuMaR DB, with 82 markers and 438 case studies) houses information on tumour markers and applicable cases as well as their specificity and sensitivity, as reported mainly from the clinical site.
5. Glyco-Disease Genes Database (GDGDB, with 80 diseases) contains information on hereditary diseases caused by mutations in glycan-related genes.

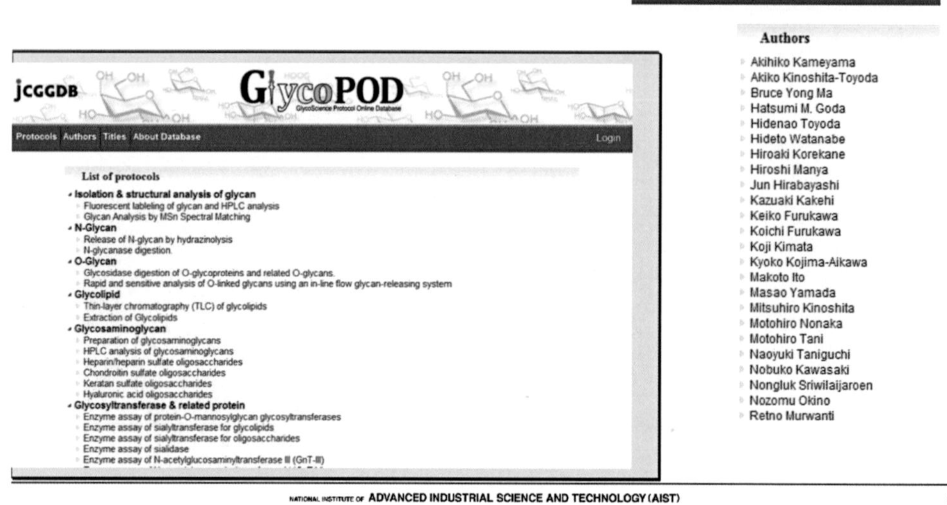

Figure 7. GlycoPOD: GlycoScience Protocol Online Database.

6. Glycoscience Protocol Online Database (GlycoPOD, with almost 200 entries, Figure 7) provides comprehensive experimental methods for studying glycoscience, including protocols for purification, analysis, fractionation, synthesis, and functional analysis.

7. Glycoside Database (Glycoside DB, with almost 70000 entries) contains information on the molecular structures of glycosides, which contain a sugar bound to another functional group via a glycosidic bond.

8. Glyco Epitope Database (GlycoEpitope, with 578 entries, Figure 8) covers information on the characteristics, usage, and applications of glyco-epitopes and antibodies.

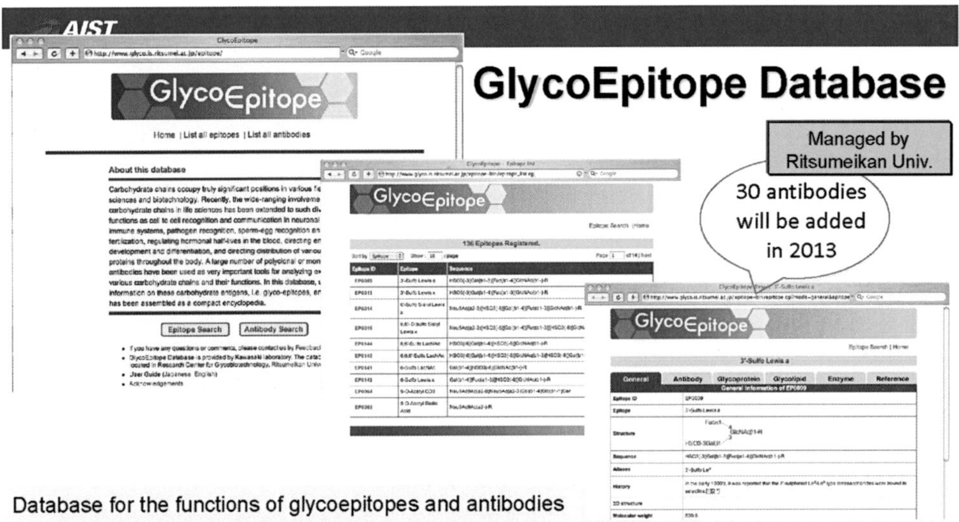

Figure 8. GlycoEpitope Database.

9. Glyco Navigation System (GlycoNavi, data on 3219 reactions) is a comprehensive database of synthetic methodology in organic chemistry, as well as information on the reagents, reaction conditions, and related literature, as well as authentic glycans and the NMR spectra of the synthetic intermediates.

10. FlyGlycoDB (with 89 entries) is a database on the phenotype of glycan-related genes in *Drosophila*, with knockdown glycan-related genes.

Mutual collaboration with the research community and data providers

The established JCGGDB was recognised as the official database of the Japan Consortium for Glycobiology and Glycotechnology, and we could build a cooperative framework with the researchers of glycoscience to integrate related databases. Moreover, we interacted extensively with the Asian researchers of glycoscience through the Asian Communications of Glycobiology and Glycotechnology (ACGG) and held regular meetings for establishing the ACGG database (Figure 9). Aiming for international standardisation, we collaborated with the world's leading databases (Figure 10), namely, UniCarbKB (Australia), GlycomeDB (Germany, US), Bacterial Carbohydrate Structure Database (BCSDB, Russia), and MonosaccharideDB (Germany), to build a glycoscience ontology in the Resource Description Framework (RDF). Simultaneously, we developed a standard notation for the glycan structure called the *Web3 Unique Representation of Carbohydrate Structures for the Semantic Web* (WURCS), and released version 1.0. We have been hosting hackathons and meetings with researchers from the US, Australia, Germany, Russia, and other countries at least once a year. We also invited researchers to the hackathons held by the Database Center for Life Science of Japan (DBCLS) for promoting collaborative research, which accelerated the development of glycoscience ontology and RDF-standardisation for each database. Thanks to the standardisation with RDF, the identifications of glycan structures are mutually linked to the data in the other standardised glycan-related databases, thus establishing the linkage between the glycan structures and information on all related proteins and lipids. This invites a broad spectrum of researchers from the fields outside glycoscience, such as molecular biology, biochemistry, and proteomics, as potential users of the repository, and thus, the number of future users may reach several millions worldwide.

AIST

Asian Communications of Glycobiology and Glycotechnology

Figure 9. Asian Communications of Glycobiology and Glycotechnology.

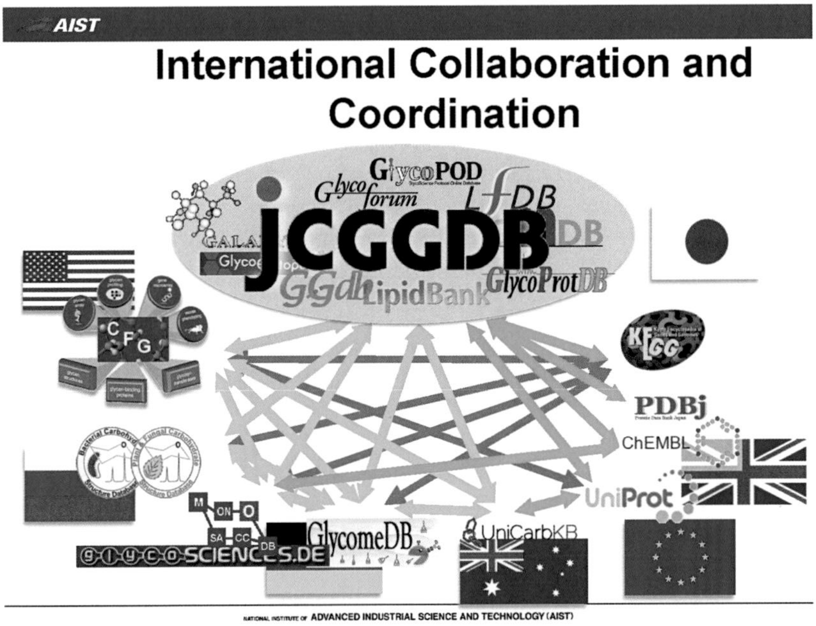

Figure 10. International Collaboration and Coordination.

Conclusion

Foreseeable tasks involved in the development of individual databases

GGDB: The analysis of glycogenes is almost complete. GGDB is currently revised accordingly.

LfDB: The contents should be increasingly enhanced. Active search for lectins naturally occurring in plants or fungi should result in the enhancement of the database. Important areas of research include modification of specifications in recombinant lectins and the analysis of lectins derived from animals, including humans.

GMDB: As there are many similar databases actively developed in other countries, collaboration and integration with such databases will be important.

GlycoProtDB: This database needs to include rapid updates reflecting the latest technical advancement. However, there has been no technology for top-down structural analysis of glycopeptides present at a small amount. In addition, it is not possible for the current technology to analyse a slight amount of a specific glycopeptide within a biological sample containing abundant non-specific molecules. A new MS-based technology to address this concern is awaited.

We have developed the lectin-microarray technique, which enabled large-scale lectin profiling of biological samples, such as various tissue specimens prepared by microdissection, cultured cell lines and their supernatants, blood, and other body fluids. Our experiments are producing large quantities of data every day; these will become available through the database in the near future.

The long-term goal in the development of glycan-related databases is the popularisation and promotion of glycoscience by increasing the recognition of glycans at a level similar to that of the gene symbols. The knowledge on glycoscience is to be collected and organised that the data can be used by researchers from the broad fields of life sciences. The technical target is knowledge sharing between glycoscience and other life science fields through a common platform, i.e., the semantic web technology. To achieve the integration of the glycoscience databases with those from other research fields via standardisation, the important base technology will be formed by standardisation of the glycan structure data and development of the international repository system. Along with the development of the base technology, we will further enhance the international coordination founded through past activities and projects.

Semantic Web Technologies Applied to Glycoscience Data to Integrate with Life Science Databases

Kiyoko F. Aoki-Kinoshita

Department of Bioinformatics, Faculty of Engineering, Soka University, 1 – 236 Tangi-machi, Hachioji, Tokyo, 192 – 8577 Japan

E-Mail: kkiyoko@soka.ac.jp

Received: 1st November 2013 / Published: 22nd December 2014

Abstract

The World Wide Web (WWW) essentially consists of web pages containing data that are linked to other resources or pages on the Internet. Therefore, to accumulate information regarding a particular carbohydrate, for example, a user would either make searches in individual databases and/or read the scientific literature and then follow various links on the Web to get relevant information. In order to overcome such tedious tasks, the Semantic Web was born from the concept of incorporating semantics, or meanings, into each data item, which is represented as a web page, or URI (Uniform Resource Identifier). Thus, a single carbohydrate structure, for example "Man_9", would be assigned a URI, and then semantics are assigned to it by preparing a dataset containing information, or annotations, about it. Each data item is annotated in the form of triples consisting of Subject, Predicate and Object. Thus, for example, "Man_9 part_of N-linked_glycan" would be a triple where Man_9 is the Subject, part_of is the Predicate and N-linked_glycan is the Object. If there was another triple, "Mannose part_of Man_9," then it can be computationally inferred that Mannose is a part of N-linked_glycan. Moreover, this triple data can be queried as a database.

This article is part of the Proceedings of the Beilstein Glyco-Bioinformatics Symposium 2013.
www.proceedings.beilstein-symposia.org

In the life sciences, many major databases have started making their data available on the Semantic Web in the form of triples, including UniProt and EBI, and the Integrated Database Project of Japan has decided to use the Semantic Web as the integrating factor among life science databases in Japan. As a part of this project, the Japanese Consortium for Glycobiology and Glycotechnology Database (JCGGDB) has called upon glycomics database providers to use Semantic Web technologies in their databases so that they can all be integrated. Thus, developers of GlycomeDB, BCSDB, MonosaccharideDB and UniCarbKB were invited to Japan to participate in a BioHackathon to learn about the Semantic Web and develop a standard by which glycomics data can be presented as triples. As an initial proof-of-concept, we were able to successfully generate on-the-fly queries across multiple databases using Semantic Web technology. In this manuscript, an introduction to the Semantic Web and these efforts to integrate glycoscience databases will be described.

Introduction

The World Wide Web (WWW), or simply the Web, has grown tremendously since it was first developed as a file server. The first version of the Web, or Web 1.0, was simply a large number of file servers that statically served files to clients through a web browser. Other than file requests, users did not have any way to interact with the server. Eventually, some web servers started providing search engines to allow users to search for data of interest. This became a part of Web 2.0, where users started becoming more active in presenting data and interacting with the Web, with blogs, tweets, social networking and of course searching.

Now it has come to be understood that even Web 2.0 is insufficient. Especially for researchers who often use the web to search for literature and existing research results, much of their time involves collecting information and integrating them to gain a better understanding of a particular research topic, for example. Thus currently, Web 3.0 has started to enable researchers to more easily search for relevant data by providing semantics to data on the Web. This is also known as the Semantic Web [1], where links are now annotated with semantic information, which the computer can potentially use to make inferences regarding relationships between available data. Realization of the Semantic Web requires the data to be annotated in the first place. Thus, in this manuscript, we describe how this is done by presenting recent work on glycomics data to transition from Web 2.0 to Web 3.0. Figure 1 illustrates the differences between Web 1.0, 2.0 and 3.0.

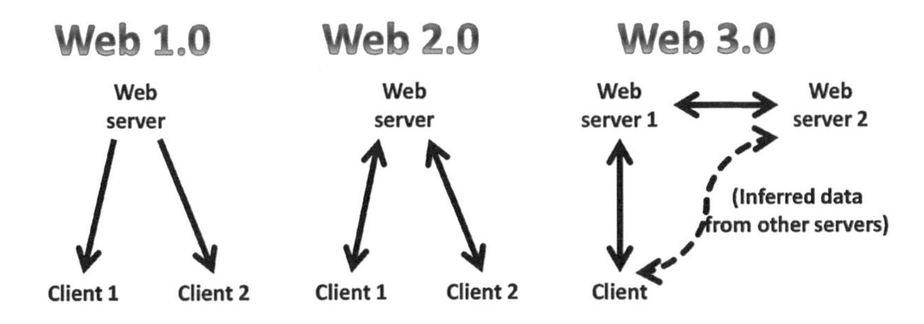

Figure 1. The transition from Web 1.0 to 2.0 to 3.0, where more interaction and eventually computational inference between data on the Web has become possible.

Background

The Semantic Web is made possible by the fact that all data is linked to one another. This network of linked data is part of what is known as the Linked Open Data (LOD), which is a network of all linked data centered on DBpedia and linked to a large variety of fields. A figure of the range of fields on the LOD can be seen at http://lod-cloud.net/. In order to add semantics to linked data, the concept of Resource Description Framework (RDF) is used, which defines data in terms of *triples* consisting of subject, predicate and object. Subjects and objects can be URLs (universal resource locators) or literals (strings). As an example, we can take the subject of the 3rd Beilstein Symposium on Glyco-Bioinformatics and define some triples about it.

Subject: 3rd Beilstein Symposium on Glyco-Bioinformatics

–Has_Theme: Discovering the Subtleties of Sugars

–Dates_Held: June 10 – 14, 2013

–At_Location: Potsdam, Germany

–Sponsor: Beilstein-Institut

Here, Has_Theme, Dates_Held, At_Location and Sponsor are predicates and "Discovering the Subtleties of Sugars," "June 10 – 14, 2013," "Potsdam, Germany" and "Beilstein-Institut" are their respective objects. We can further define more triples on the subject of Potsdam, as follows.

Subject: Potsdam, Germany

–In_State: Brandenburg

–Has_Population: 158,902

–Has_Website: http://www.potsdam.de

Note that Potsdam in the previous subject can then be linked to this subject, as illustrated in Figure 2.

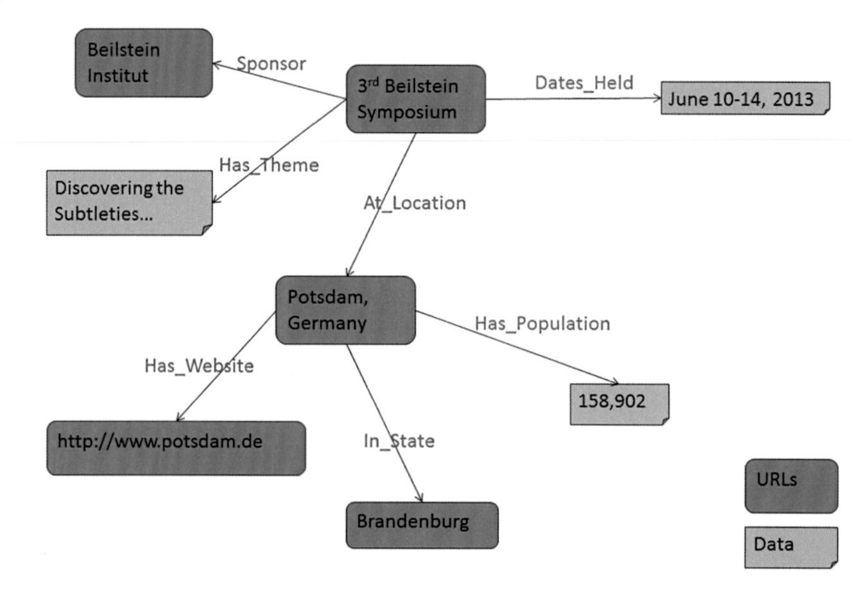

Figure 2. A network graph illustrating the linked relationships between the triples described as examples in the text.

Predicates assign meaning to data, and they are usually defined in an *ontology*. An ontology is traditionally defined as "a systematic account of existence", and more recently has taken on the following definition: "the hierarchical structuring of knowledge about things by subcategorizing them according to their essential qualities." The Gene Ontology (GO) is a well-known ontology of hierarchically organized terms that can be used to annotation gene information [2]. Using GO, genes can be annotated with consistent terms, making it easier for users to search for genes with similar function or cellular localization, etc. The Semantic Web relies heavily on ontologies in order to add meaning to data and their relationships in a standardized manner.

Once data is annotated with triples and stored in RDF format, the data are entered into a special RDF database called a triplestore and made available to queries at a special URI which is called a SPARQL endpoint. The data can then be queried using the SPARQL query language for RDF data. Once the data is stored in a triplestore, software can be used to make inferences on the data. For example, using the example described previously, computers may make an inference on the Beilstein Symposium as follows: "The 3rd Beilstein Symposium on Glyco-Bioinformatics will be held in the German state of Brandenburg." In this example, note that the data never directly linked the information about the 3rd Beilstein Symposium with Brandenburg. However, the computer can make the inference through the direct links with Potsdam. To extend this example in terms of glycosciences, glycan structures, for example, can be linked with information about the related glycoproteins, glycolipids, diseases, gene information, etc. Then computers would be able to make inferences and provide clues to researchers about glycan function through these links. Such work is currently done by all researchers when surveying the literature about their target glycan or protein of interest. With the aid of the Semantic Web, however, the time and effort required for such literature surveys can be greatly decreased.

Next, we focus on the technical aspects that must be addressed in order to transfer the glycomics data stored in the currently publicly available glyco-databases onto the Semantic Web. Databases such as JCGGDB, GlycomeDB, BCSDB, UniCarbKB and MonosaccharideDB were represented by the corresponding database developers, who gathered at BioHackathons held in Japan and China in 2012 and 2013, respectively. We will describe the results of these hackathons in this manuscript.

Methods

GlycoRDF

The National Bioscience Database Center (NBDC) and Database Center for Life Science (DBCLS) in Japan have held BioHackathons annually since 2008. BioHackathon stands for Biology + Hacking + Marathon, where programmers come together to develop software together for a few days (usually a week). The 5th Annual DBCLS BioHackathon was held in Toyama city, Japan, from September 2 – 7, 2012. Developers of major glycan databases worldwide gathered together in Toyama to learn about the Semantic Web and develop RDF versions of their respective databases. The databases involved are GlycomeDB [3], UniCarbKB [4], Japan Consortium for Glycobiology and Glycotechnology DataBase (JCGGDB) (http://jcggdb.jp/index_en.html), UniProt, GlycoEpitope (http://www.glyco.is.ritsumei.ac.jp/epitope2/), MonosaccharideDB (http://www.monosaccharidedb.org) and Bacterial Carbohydrate Structure Database (BCSDB) [5]. Their main focus at the BioHackathon was to ensure consistency between the predicates and the links used in each RDF data set.

SPARQL

After preliminary RDF data were generated from each participant, all of the data was stored in a local triplestore so as to test SPARQL queries on the integrated data. Several queries were tested, which are summarized in Table 1.

Table 1. List of queries tested at the 5th BioHackathon to assess useability of RDF on glycoscience data.

	Description	Databases Involved	Input: Output
Query 1	Obtain UniProt protein IDs for glycan structures in JCGGDB.	JCGGDB, GlycomeDB, UniCarbKB, UniProt	JCGGDB ID: UniProt ID
Query 2	Retrieve the glycan structures involved with lectins across all databases.	Lectin Frontier Database (LfDB) of JCGGDB, GlycomeDB	Lectin data: Glycan structure data
Query 3	Search for the carrier proteins of glycan epitopes.	GlycoEpitope, UniProt, GlycoProtDB (Kaji, et al., 2012) of JCGGDB	GlycoEpitope data: NCBI RefSeq IDs referenced from GlycoProtDB

RESULTS

Each of the queries tested during the 5th BioHackathon could be successfully carried out on our preliminary RDF data set. Although each query involved multiple databases, a single SPARQL query could be written to obtain the requested data in one run. That is, each query could be made with one SPARQL statement, whereas users would be required to access several different databases to obtain the same information over the traditional Web. Thus, the power of the Semantic Web could be demonstrated. As a result, the glyco-database developers could also get a better understanding of what is involved in developing data for the Semantic Web. Detailed results have been published in [6].

The results of the 5th BioHackathon enabled glyco-database developers to get an idea for the potential of the Semantic Web. However, the development process of RDF data generation from each database and generating SPARQL queries also showed us the importance of a standardizing ontology of glyco-data. In particular, the Semantic Web enables computers to make inferences on the RDF data, which requires a consistent ontology to be defined in the first place. Thus, the developers at the 5th Hackathon decided to gather again at a glyco-hackathon, which was held during the GLYCO22 meeting in Dalian, China, June 23 – 28, 2013 [7]. This meeting was sponsored by the Advanced Institute for Science and Technology (AIST) and was focused on the development of a standardized ontology for glycan structures and related data such as publications, experimental procedures and biological samples. The results of this work will allow database developers to generate RDF using consistent predicates on the appropriate subject classes and for the relevant object classes. Moreover, recently the EBI has announced the availability of their life science data on the Semantic Web [8]. Thus, in addition to UniProt, more biological data are now available to be linked.

Conclusion and Future Perspectives

The next steps will include the development of more intuitive user interfaces to query and make inferences on the glycomics and related life science data that are slowly become a part of the Semantic Web. As one of the informatics aims proposed in the NAS report on glycosciences [9], the next five and ten years will see an increased availability of glycomics data on the web along with better computational tools. It can be expected that the Semantics Web technologies will serve as key contributors to this progress.

Acknowledgements

This manuscript presents work that was made possible by the glyco-database developers who participated in the BioHackathons in Toyama and Dalian: Jerven Bolleman (UniProt), Matthew P. Campbell (UniCarbKB), Shin Kawano and Jin-Dong Kim (DBCLS), Thomas Luetteke (MonosaccharideDB), Shujiro Okuda (GlycoEpitope), Rene Ranzinger (GlycomeDB), Hiro Sawaki, Daisuke Shinmachi and Hisashi Narimatsu (JCGGDB at AIST), and Philip Toukach (BCSDB). This work has also been supported by National Bioscience Database Center (NBDC) of Japan Science and Technology Agency (JST), National Institute of Advanced Industrial Science and Technology (AIST) in Japan, and the Database Center for Life Science (DBCLS) in Japan. The developers also recognize the invaluable contributions from the community and those efforts to curate and share structural and experimental data collections.

References

[1] Berners-Lee, T., Hendler, J. and Lassila, O. (2001) The Semantic Web. *Scientific American*. pp. 29 – 37.

[2] Blake, J.A., *et al.* (2013) Gene Ontology annotations and resources. *Nucleic Acids Res.* **41**:D530 – 535.
http://dx.doi.org/10.1093/nar/gks1050.

[3] Ranzinger, R., *et al.* (2011) GlycomeDB – a unified database for carbohydrate structures. *Nucleic Acids Res.* **39**:D373 – 376.
http://dx.doi.org/10.1093/nar/gkq1014.

[4] Campbell, M.P., *et al.* (2011) UniCarbKB: putting the pieces together for glycomics research. *Proteomics* **11**:4117 – 4121.
http://dx.doi.org/10.1002/pmic.201100302.

[5] Toukach, P., *et al.* (2007) Sharing of worldwide distributed carbohydrate-related digital resources: online connection of the Bacterial Carbohydrate Structure DataBase and GLYCOSCIENCES.de. *Nucleic Acids Res.* **35**:D280–286. http://dx.doi.org/10.1093/nar/gkl883.

[6] Aoki-Kinoshita, K.F., *et al.* (2013) Introducing glycomics data into the Semantic Web. *Journal of Biomedical Semantics* 2013, **4**:39. http://dx.doi.org/10.1186/2041–1480–4-39.

[7] Aoki-Kinoshita, K.F., *et al.* (2013) The Fifth ACGG-DB Meeting Report: Towards an international glycan structure repository. *Glycobiology*, **23** (12): 1422–1424. http://dx.doi.org/10.1093/glycob/cwt084.

[8] EMBL-EBI (2013) Bioinformatics embraces Semantic Web technologies.

[9] Committee on Assessing the Importance and Impact of Glycomics and Glycosciences, N.R.C.U. (2012) *Transforming Glycoscience: A Roadmap for the Future.* National Academies Press Washington, DC, USA.

Utilising the Carbohydrate Fragmentation Database UniCarb-DB for Glyco Research

Catherine A. Hayes, Sarah A. Flowers, Liaqat Ali, Samah M. A. Issa, Chunsheng Jin and Niclas G. Karlsson*

Medical Biochemistry, Sahlgrenska Academy, University of Gothenburg, Box 440 405 30 Gothenburg, Sweden

E-Mail: *niclas.karlsson@medkem.gu.se

Received: 5th December 2013 / Published: 22nd December 2014

Abstract

UniCarb-DB is an experimental database consisting of structural information of *O*-linked and *N*-linked oligosaccharides and associated LC-MS/MS fragmentation data. This report illustrates how the database can be useful for future software development for the interpretation of glycomic mass spectrometric data. The information about rich fragmentation spectra generated for *O*-linked oligosaccharides in negative ion mode allows matching with candidate spectra in the database for accurate assignment of oligosaccharide sequence and in some cases, linkage information. Furthermore, peak matching can be used for selected *m/z* regions to identify sites of sulfation and identification of the sequence of the neutral oligosaccharide backbone of sialylated structures. The reproducibility of the fragmentation patterns of oligosaccharides present in the database suggests that targeted mass spectrometric approaches can be developed for glycomic discovery and validation, using such methods as multiple/selected reaction monitoring (MRM/SRM).

This article is part of the Proceedings of the Beilstein Glyco-Bioinformatics Symposium 2013.
www.proceedings.beilstein-symposia.org

Introduction

The lack of structural databases for glycomic research is a key obstacle for successful high throughput glycomic analysis to identify the role of glycosylation in life science. Progress has been made in the last couple of years. While the pioneering work to establish CarbBank [1] as a universal carbohydrate databank was discontinued in the 90s, other groups have now taken up the challenge that will allow glycomic structural databases to grow again. For instance, the UniCarbKB project [2] is building on the framework established for the GlycoSuite database [3], capturing structural information of glycoproteins from the scientific literature. Another example is GlycomeDB [4] which aims to amalgamate the information about carbohydrate structures available in existing databases into one depository. Publically available experimental glycomic databases, where structural data are associated with characterisation information, have been established for NMR [5] HPLC [6] and LC-MS^2 data [7]. The latter, named UniCarb-DB, is based on *de novo* sequencing of both *N*-linked and *O*-linked oligosaccharides primarily based on fragmentation analysis in negative ion mode. To date, the database contains over 500 MS/MS spectra representing 416 uniquely defined structures. The association of structures with MS/MS spectra allows the database to be utilised in downstream glycomic MS based analysis to match sample spectra with consensus spectra from the database, and provides a pathway to design targeted MS experiments of predetermined components in SRM. An SRM approach has been shown to be an effective pathway to monitor low abundant sulfated oligosaccharides present in inflammatory diseases [8].

Methods

MS analysis of O-linked oligosaccharides

Ion trap LC-MS/MS spectra of neutral and sialylated structures were generated from oligosaccharides released from human gastric mucins as described [9]. LTQ orbitrap (Thermo Electron, San Jose, CA, US) Higher energy Collision induced Dissociation (HCD) was performed on *O*-linked oligosaccharides released from porcine gastric mucin (PGM) (Sigma, St Lois, MO, US) or human salivary MUC 5B [10]. The Obritrap was calibrated and tuned in negative ion mode with the manufacturer's standard mixture. Released *O*-linked oligosaccharides were fragmented in the HCD cell with a normalised energy of 90. The eluate from a graphitized carbon capillary HPLC column (produced and run as described [9]) system was introduced using the standard ion max source. Ions were generated and focused using an ESI voltage of – 3,750 V, a sheath gas flow of 20 L/min, and a capillary temperature of 250 °C. MS data acquisition was carried out with the LTQ-Orbitrap mass spectrometer scanning in negative ion mode over *m/z* 700 – 1650, with a resolution of 30,000 at *m/z* 400. This was followed by data dependent MS^2 scans of the three most abundant ions in each scan

(2 microscans, maximum 500 ms, target value of 100,000). The signal threshold for MS^2 was set to 5,000 counts with an isolation window of 3 Da, and an activation time of 10 ms was used.

Matching of fragmentation data with spectral library

MS/MS peak lists from fragmentation of oligosaccharides were centroided using Xcalibur and compared with the spectra in UniCarb-DB. The R package OrgMassSpecR[3] was used, specifically the SpectrumSimilarity function. This function takes two spectra and compares them. The output is a similarity score based on the normal dot product.

A subset library of 231 MS/MS spectra from UniCarb-DB was created for matching with the neutral and sialylated structures. A test-set of seven unknown spectra was used. This set included neutral and sialylated structures. Matching of the human gastric neutral and sialylated MS/MS spectra with the database against the 231-structure database was performed using a precursor mass filter, t = 0.25 and b = 5, where t is the mass tolerance (*m/z*) used to align the spectra, and b is the baseline threshold for peak identification (expressed as % of max. intensity). Sulfated oligosaccharides from PGM and MUC5B was matched with a smaller spectral library consisting of sulfated *N*-acetyllactosamine with known sulfate position (generous gift from James Paulson, Scripps Research Institute, La Jolla, CA, USA).

Results

MS fragmentation of neutral, sialylated and sulfated oligosaccharides provides different information that needs to be considered when utilising UniCarb-DB fragmentation data

The chemical nature of *O*-linked oligosaccharide subclasses i.e., neutral, sialylated and sulfated oligosaccharides, influences their fragmentation in negative ion mode (Figure 1). The negative charge introduced on neutral oligosaccharides can be spread throughout the molecules by removing a proton from either of the omnipresent alcohol groups throughout the oligosaccharide chain. The close proximity of the charge to the neighbouring carbons allows the fragmentation to progress via both charge remote and charge induced fragmentation [11]. The acidic sulfate and sialic acid residues, on the other hand, allow the charge to be specifically localised to these residues. The close proximity of the carboxyl group of sialic acid residues to a glycosidic bond promotes loss of the sialic acid as the main fragmentation pathway as seen in Figure 1 middle panel, where the fragment ion of *m/z* 1040 is the dominating ion. Further fragmentation after removal of the sialic acid then progresses similarly to fragmentation of neutral oligosaccharides, that is, according to a decentralised charge carried by a deprotonated poly-ol. For sulfated oligosaccharides, the sulfate group appears to be more stable during the fragmentation process compared to sialic acid. This stability promotes charge–remote fragmentation and progresses mainly *via* glycosidic Y and B cleavages. The knowledge of how different oligosaccharide subclasses fragment will impact on how the database can be used for glycol research.

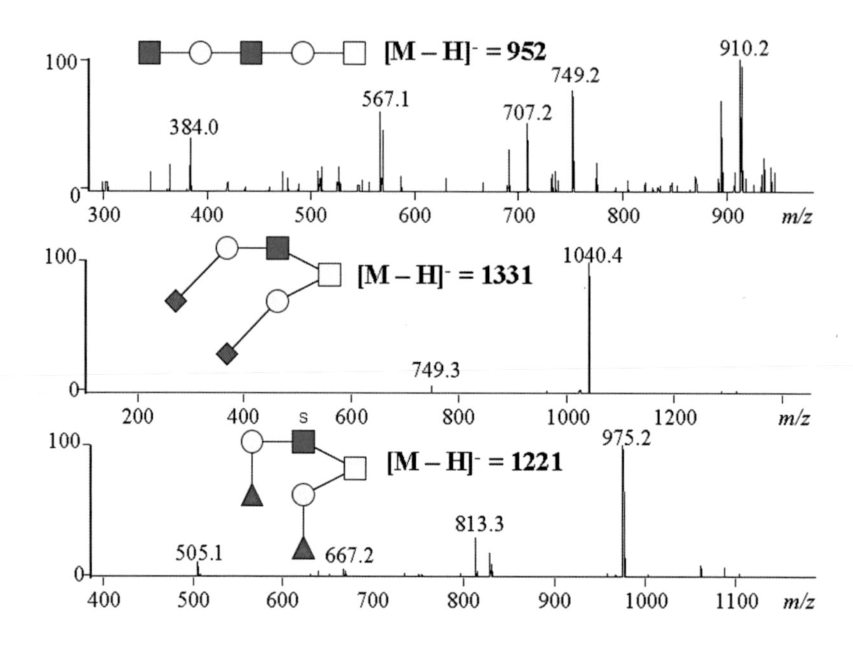

Figure 1. Illustration of the differences between the fragmentation in negative ion CID of neutral (upper panel), sialylated (middle panel) and sulfated (lower panel) *O*-linked oligosaccharides.

Matching fragment spectra of neutral O-linked oligosaccharides with UniCarb-DB spectral library

Peak matching of *m/z* and intensity data of unidentified components with a spectral library has become one of the most successful approaches for mass spectrometry, originally implemented for small molecules and GC-MS with electron impact (EI) fragmentation. This type of fragmentation is very reproducible between different instruments. Hence, a database of small molecule standards and associated fragmentation has been developed and currently the National Institute of Standards and Technology contains over 200,000 EI spectra (http://www.nist.gov/srd/nist1a.cfm). Collision induced dissociation (CID) used in combination with ES and MALDI is less standardized and factors such as charge state, collision gas, collision energy and type of mass spectrometer (triple quadrupole instruments, quadrupole-time of flight or ion trap) will influence the fragment spectra. Hence, the current initiative for Minimum Information Required for A Glycomics Experiment (MIRAGE) [12] is important for any glycomic spectral MS library containing all this meta data, to be able to compare data generated under similar conditions. In Figure 2, MS/MS spectra from ion trap CID of two isomeric *O*-linked oligosaccharides from human gastric mucins with the $[M\text{–}H]^-$-ions of *m/z* 749 is compared with the UniCarb-DB spectral library (ion trap data) using the dot-product algorithm. For the structure identified as Galβ1-3(Galβ1-4GlcNAcβ1-6)GalNAcol (left panel) the dot product scoring indicates almost identical spectra in regards of intensities of all *m/z* values. The isomer matched in the right panel of Figure 2 showed to be less consistent spectra with its match. Closer examination of this spectrum and the meta data showed that it was generated under similar conditions and is probably not the reason for this difference. Hence, the data suggests that the sequence of the matched unknown has a similar linear sequence Hex-HexNAc-Hex-HexNAcol, but differs from its match either on linkage configuration, linkage type and/or difference between isomeric monosaccharide units. The two examples illustrate the caution in utilising the database for anything else but to identify candidate structures for the oligosaccharide of interest. This also suggests that a statistically significant match does not have to be found in order to get insight into the structure of unidentified oligosaccharides.

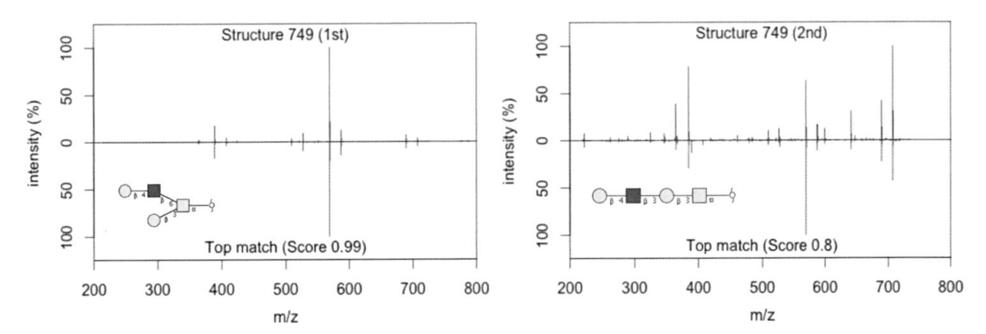

Figure 2. Results of the two neutral *O*-linked oligosaccharides queried against the database. Shown are the top-to-tail plots of the query (top) versus the best match (bottom) MS/MS along with the score (based on normal dot product).

Matching fragment spectra of sialylated O-linked oligosaccharides with spectral library

The fragmentation observed for sialylated structures, with the predominant loss of sialic acid and additional low intense fragments from fragmentation of the neutral backbone, makes the localisation of the sialic acid within the molecule difficult. We have previously devised a method using sialidase treatment in combination with LC-MS^2 to generate high quality fragmentation of its neutral counterpart. An alternative approach would be to utilise the inherent ability of the mass spectrometer to perform desialylation during the fragmentation, and acknowledge that there is little information in the spectra that tells where the sialic acid is located. This involves the partial matching of the sialylated fragment data against a spectral library containing desialylated (neutral) versions of the structures. In Figure 3, the fragment spectra of the neutral branched core 2 structure provides a better match to the monosialylated [M–H] ion of *m/z* 1040 than the neutral linear core 1 structure. This is possible by using the neutral backbone mass-range for the matching and excluding the dominating Y fragment of *m/z* 749 (loss of sialic acid) (Figure 1B).

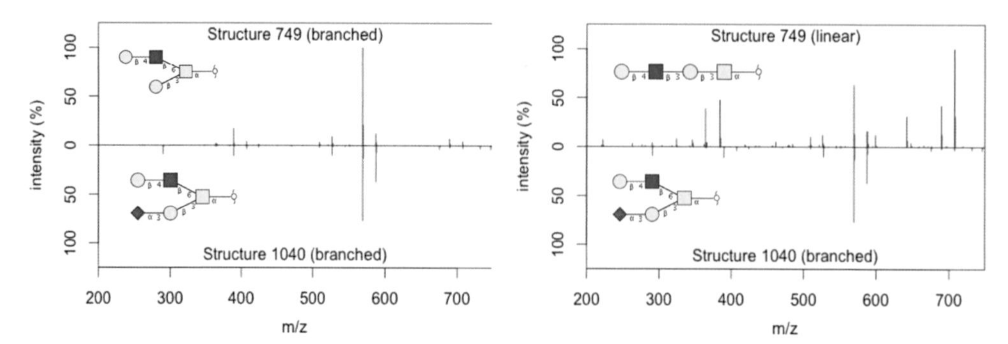

Figure 3. Results of comparing the neutral back-bone region of a sialylated structure with [M–H] ion of *m/z* 1040 structure to two neutral MS/MS spectra. This enabled the identification of the neutral nature of this particular structure; score of branched core 2 structure of [M–H] of *m/z* 749 to 1040 was 0.94 (left panel), whereas it was only 0.4 for the linear core 1 structure (right panel).

The partial matching of sialylated spectra with neutral spectra from the UniCarb-DB library illustrates how the fragmentation can be used intelligently for querying the library. However, care must be taken not to over interpret results. In this particular case the isomers NeuAcα2-3Galβ1-3(Galβ1-4GlcNAcβ1-6)GalNAcol and Galβ1-3(NeuAcα2-3Galβ1-4GlcNAcβ1-6)GalNAcol, [M–H] ion of *m/z* 1040, display almost identical MS/MS spectra, but the structure of the two pairs of can be distinguished by the difference in LC-MS retention time [13].

Matching fragment spectra of sulfated O-linked oligosaccharides with spectral library

Knowledge of the structure of sulfated oligosaccharides is still scarce, and only a limited number of ms/ms are recorded in UniCarb-DB. We set out to investigate if we could build a small library of known sulfation positions available on mucin type oligosaccharides, and explored if we can predict the type of sulfation present (6-linked GlcNAc or 3-linked Gal). The sulfate group of the oligosaccharide will carry the negative charge in the electrospray and promote charge remote fragmentation. This provides cross ring fragments in the low molecular mass region (*m/z* < 350) from single and multiple (internal) fragmentations using HCD fragmentation. Figure 4 (left panel) shows the matching of the low mass region of a sulfated oligosaccharide with [M–H] ion of *m/z* 1324 (HSO_3-$Fuc_2Hex_2HexNAc_2$HexNAcol) from human salivary MUC5B with the MS/MS of a sulfated standard HSO_3-3Galβ1-4GlcNAc. The data generated by HCD fragmentation [14] is clearly distinguished from the HCD fragmentation of the 6 sulfated GlcNAc from porcine gastric mucin ([M–H] ion of *m/z* 1178, HSO_3-$Fuc_1Gal_2HexNAc_2$HexNAcol) and the standard Galβ1-4(HSO_3-6) GlcNAc (Figure 4 right panel).

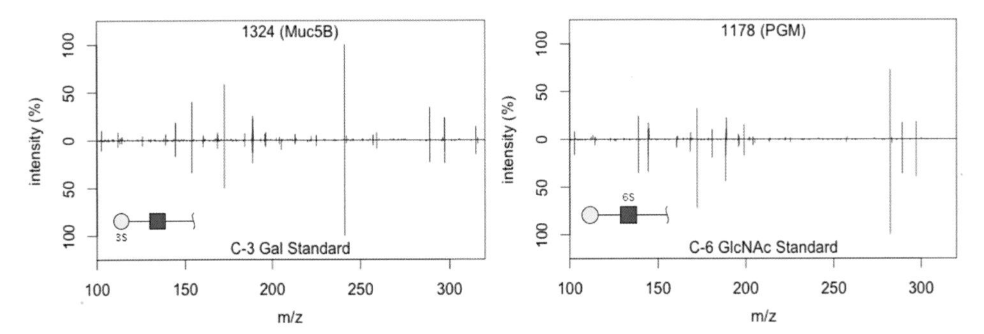

Figure 4. Top two matching structures from sulfated human salivary MUC5B structure with the [M–H] ion of *m/z* 1324 (left top) with in the C-3 linked sulfated Gal standard in the *m/z* 100 – 320 region (0.89) (left bottom) and the PGM structure with the [M–H] ion of *m/z* 1178 (right top) with the C-6 linked sulfated GlcNAc standard (right bottom) with the C-6 linked GlcNAc standard (0.85).

The charge remote fragmentation seen for sulfated oligosaccharides makes it possible to identify the site of sulfation without actually having the identical oligosaccharide available in the spectral library. The nature of the fragmentation distinguishing the neutral, sialylated and sulfated structures allows a spectral library, such as UniCarb-DB, to be utilised intelligently to identify oligosaccharide sequence, including at times, linkage information within a sample analysed by LC-MS^2.

Target driven glycomic-Selective Reaction Monitoring (SRM)

ELISA and western blotting methods for individual proteins and protein families are widely used in life science and routinely for clinical diagnostics. In a similar fashion, the monitoring of individual oligosaccharides and/or oligosaccharide families to address biological questions would be useful in glycobiology. SRM for quantification of peptides have been shown to be a very efficient way for quantification in targeted proteomics [15]. This methodology requires insight into the fragmentation of the molecule to accurately determine transitions that would be used for monitoring protein signature peptides. In proteomics these transitions can be verified by standard synthetic peptides. However, commercial standards are not available for most *O*-linked oligosaccharides.

SRM using triple quadrupole mass spectrometry in negative ion mode and the excellent separation using graphitised carbon would provide a pathway for setting up targeted glycomics. This approach has been utilised to show how sulfation influences the extension of core 1 *O*-linked glycosylation in inflammation [8]. In Table 1 it can be seen that the data in UniCarb-DB could be used to design this type of targeted approach for investigating relationship between families of *O*-linked oligosaccharides. The table shows the 11 *O*-linked oligosaccharides that would be found in plasma (core 1 and core 2 oligosaccharides). Without access to standards the table allows access to transitions for SRM from the UniCarb-DB spectra library, since all these structures have been recorded in the database with their associated negative ion mode MS/MS spectra. Most of the glycans provide specific fragmentation/precursor ion transition pairs that would allow them to be specifically identified. In regards to some of the isomeric structures like the pairs NeuAcα2-3Galβ1-3GalNAcol/ Galβ1-3(NeuAcα2-6)GalNAcol with [M–H] of *m/z* 675 and the NeuAcα2-3Galβ1-3 (Galβ1-4GlcNAcβ1-6)GalNAcol/Galβ1-3(NeuAcα2-3Galβ1-4GlcNAcβ1-6)GalNAcol with [M–H] of *m/z* 1040, within each isomeric pair the most intense transitions would also have the same mass. Since these structures are clearly separated by chromatography, their identity may not only be based on the transitions and their individual intensity relationship but also on the retention time.

$[M-H]^-$	Structure	*Glycan ID*
384	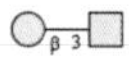	*367*
675	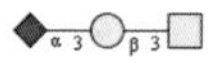	*1141*
675	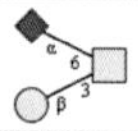	*572*
966	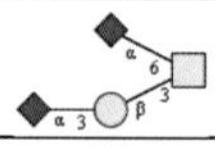	*1142*
587	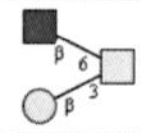	*568*
749	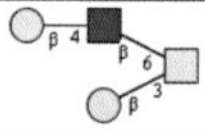	*581*
1040	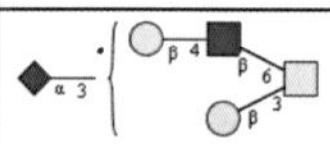	*3696*
1331	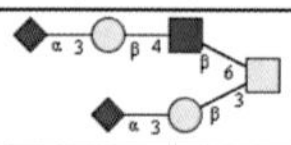	*2250*
829	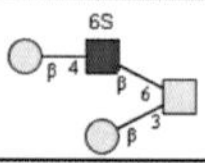	*28901*
1120	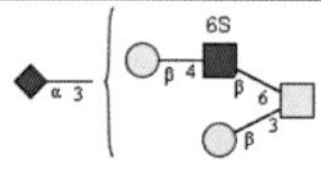	*3239*
1411	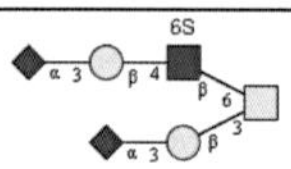	*9673*

Table 1. *O*-linked oligosaccharides present in human serum/plasma and their ID as recorded in UniCarb-DB

Discussion

Spectral libraries of oligosaccharides (such as UniCarb-DB) can be useful for glycomic analysis. The matching of query fragment spectra with the spectra recorded in the experimentally generated spectral library facilitates a high-throughput glycomic identification workflow. With the increasing number of spectra generated, this process will be easier, and the focus of *O*-linked glycomic discovery will be able to shift away from structural characterisation and focus on structural verification and quantitative aspects related to regulation of various glyco-epitopes. Spectral matching, used intelligently, can also aid in the structural assignment of novel structures, as was illustrated by the identification of the sulfate position of unknown sulfated oligosaccharides in PGM and MUC5B. The UniCarb-DB spectral library is currently primarily based on CID in negative ion mode using ion trap fragmentation. As this database becomes an acknowledged site for storage of oligosaccharide fragmentation spectra in the research community, we hope to include other fragmentation methods, ion modes, mass spectrometer types and various modification (various adducts, reducing end, permethylation, peracetylation and other ways of modification such as methyl ester formation of sialylation).

Going beyond discovery into biomarker screening, we expect that the spectral library would be useful in the design of SRM experiments for *O*-linked oligosaccharides (Figure 4). This would allow hypothesis driven glycomic research, where certain structures or family of structures can be targeted. Beyond this, a structural spectral library would be a prerequisite for querying very large amounts of LC-MS based oligosaccharide fragment data, as for instance can be generated with novel type of broad range simultaneous precursor fragmentations [16]. As the synthetic generation of all of the complex *O*-linked structures present in mammalia is still in the future, there is no choice but to use the natural sources to generate a valuable resource such as UniCarb-DB. To be successful, we are convinced that this resource needs the contribution from the whole glycomic MS society. In addition to the information about SRM transitions, it also provides information about biological sources from where individual structures can prepared. Access to sources where oligosaccharides of interest are present is important for the MS optimization of transitions.

Acknowledgements

This work was supported by the Swedish Research Council (621 – 2010 – 5322) and the Swedish Foundation for International Cooperation in Research and Higher Education. The mass spectrometer was obtained by a grant from the Swedish Research Council (342 – 2004 – 4434) and from Knut and Alice Wallenberg's Foundation (KAW2007.0118).

References

[1] Doubet, S., Bock, K., Smith, D., Darvill, A., Albersheim, P. (1989) The Complex Carbohydrate Structure Database. *Trends in Biochemical Sciences* **14**:475 – 477. doi: 10.1016/0968-0004(89)90175-8

[2] Campbell, M.P., Hayes, C.A., Struwe, W.B., Wilkins, M.R., Aoki-Kinoshita, K. F., Harvey, D.J., Rudd, P.M., Kolarich, D., Lisacek, F., Karlsson, N.G., Packer, N.H. (2011) UniCarbKB: Putting the pieces together for glycomics research. *Proteomics* **11**:4117 – 4121. doi: 10.1002/pmic.201100302

[3] Cooper, C.A., Harrison, M.J., Wilkins, M.R., Packer, N.H. (2001) GlycoSuiteDB: a new curated relational database of glycoprotein glycan structures and their biological sources. *Nucleic Acids Research* **29**:332 – 335. doi: 10.1093/nar/29.1.332

[4] Ranzinger, R., Frank, M., von der Lieth, C.-W., Herget, S. (2009) Glycome-DB.org: a portal for querying across the digital world of carbohydrate sequences. *Glycobiology* **19**:1563 – 1567. doi: 10.1093/glycob/cwp137

[5] Lundborg, M., Widmalm, G. (2011) Structural Analysis of Glycans by NMR Chemical Shift Prediction. *Analytical Chemistry* **83**:1514 – 1517. doi: 10.1021/ac1032534

[6] Royle, L., Campbell, M.P., Radcliffe, C.M., White, D.M., Harvey, D.J., Abrahams, J.L., Kim, Y.-G., Henry, G.W., Shadick, N.A., Weinblatt, M.E., Lee, D.M., Rudd, P.M., Dwek, R.A. (2008) HPLC-based analysis of serum *N*-glycans on a 96-well plate platform with dedicated database software. *Analytical Biochemistry* **376**:1 – 12. doi: 10.1016/j.ab.2007.12.012

[7] Hayes, C.A., Karlsson, N.G., Struwe, W.B., Lisacek, F., Rudd, P.M., Packer, N.H., Campbell, M.P. (2011) UniCarb-DB: a database resource for glycomic discovery. *Bioinformatics* **27**:1343 – 1344. doi: 10.1093/bioinformatics/btr137

[8] Flowers, S.A., Ali, L., Lane, C.S., Olin, M., Karlsson, N.G. (2013) Selected Reaction Monitoring to Differentiate and Relatively Quantitate Isomers of Sulfated and Unsulfated Core 1 *O*-Glycans from Salivary MUC7 Protein in Rheumatoid Arthritis. *Molecular & Cellular Proteomics* **12**:921 – 931. doi: 10.1074/mcp.M113.028878

[9] Kenny, D.T., Skoog, E.C., Lindén, S.K., Struwe, W.B. Rudd, P.M., Karlsson, N.G. (2012) Presence of Terminal *N*-acetylgalactosamineβ1-4*N*-acetylglucosamine Residues on *O*-linked Oligosaccharides from Gastric MUC5AC: Involvement in *Helicobacter pylori* Colonization? *Glycobiology* **22**:1077 – 1785.
doi: 10.1093/glycob/cws076

[10] Thomsson, K.A., Schulz, B.L., Packer, N.H., Karlsson, N.G. (2005) MUC5B glycosylation in human saliva reflects blood group and secretor status. *Glycobiology* **15**:791 – 804.
doi: 10.1093/glycob/cwi059

[11] Doohan, R.A., Hayes, C.A., Harhen, B., Karlsson, N.G. (2011) Negative ion CID fragmentation of *O*-linked oligosaccharide aldoses-charge induced and charge remote fragmentation. *Journal of The American Society for Mass Spectrometry* **22**(6):1052 – 1062.
doi: 10.1007/s13361-011-0102-3

[12] Kolarich, D., Rapp, E., Struwe, W.B., Haslam, S.M., Zaia, J., McBride, R., Agravat, M.P., Campbell, M.P., Kato, M., Ranzinger, R., Kettner, C., York, W.S. (2013) The Minimum Information Required for A Glycomics Experiment (MIRAGE) Project: Improving the standards for reporting Mass spectrometry-based Glycoanalytic Data. *Molecular & Cellular Proteomics* **12**:991 – 995.
doi: 10.1074/mcp.O112.026492

[13] Karlsson, N.G., McGuckin, M.A. (2012) *O*-Linked glycome and proteome of high-molecular-mass proteins in human ovarian cancer ascites: Identification of sulfation, disialic acid and *O*-linked fucose. *Glycobiology* **22**:918 – 929.
doi: 10.1093/glycob/cws060

[14] Olsen, J.V., Macek, B., Lange, O., Makarov, A., Horning, S., Mann, M. (2007) Higher-energy C-trap dissociation for peptide modification analysis. *Nature Methods* **4**:709 – 12.
doi: 10.1038/nmeth1060

[15] Gallien, S., Duriez, E., Domon, B. (2011) Selected reaction monitoring applied to proteomics. *Journal of Mass Spectrometry* **46**:298 – 312.
doi: 10.1002/jms.1895

[16] Law, K.P., Lim, Y.P. (2013) Recent advances in mass spectrometry: data independent analysis and hyper reaction monitoring. *Expert Review of Proteomics* **10**:551 – 566.
doi: 10.1586/14789450.2013.858022

UniCarbKB: First Year Report Card

Matthew P. Campbell[1], Robyn Peterson[1], Elisabeth Gasteiger[2], Jingyu Zhang[1], Yukie Akune[3], Jodie L. Abrahams[1], Julien Mariethoz[4], Catherine A. Hayes[5], Daniel Kolarich[6], Niclas G. Karlsson[5], Kiyoko F. Aoki-Kinoshita[3], Frederique Lisacek[4], Nicolle H. Packer[1,*]

[1]Biomolecular Frontiers Research Centre, Macquarie University, North Ryde, Sydney, NSW 2109, Australia,

[2]Swiss-Prot Group, Swiss Institute of Bioinformatics, CH-1211 Geneva, Switzerland

[3]Department of Bioinformatics, Faculty of Engineering, Soka University, 1 – 236 Tangi-machi, Hachioji, Tokyo, Japan

[4]Proteome Informatics Group, Swiss Institute of Bioinformatics, CH-1211 Geneva, Switzerland

[5]Department of Medical Biochemistry and Cell Biology, University of Gothenburg, S-40 530 Gothenburg, Sweden

[6]Department of Biomolecular Systems, Max Planck Institute of Colloids and Interfaces, 14424 Potsdam, Germany

E-Mail: *nicki.packer@mq.edu.au

Received: 19th December 2013 / Published: 22nd December 2014

Abstract

At the Beilstein Workshop on Glycoinformatics in 2011 we introduced UniCarbKB as an international initiative that aims to collect and distribute resources and practices from glycobiology practitioners to the whole biological research community. The mission was, and still is, to provide a comprehensive, high quality catalogue of information on carbohydrates, and to continue efforts to advance the interpretation of captured data through the development of novel data analysis methods and algorithms for the efficient representation and mining of large experimental data sets.

This article is part of the Proceedings of the Beilstein Glyco-Bioinformatics Symposium 2013.
www.proceedings.beilstein-symposia.org

Here, we report the progress we have made on establishing the infrastructure and content of the fledgling UniCarbKB. This will include our current work on the integration, into the publically available UniCarbKB portal, of data from UniCarb-DB, GlycoSuiteDB, GlycoBase, EUROCarbDB, SugarBind and PubChem. In the future it is hoped that the UniCarbKB knowledgebase, based on a central database of curated glycan structures, will become the key resource of quality information for glycobiology research.

Introduction

The NIH report Transforming Glycoscience: A Roadmap for the Future [1] identified that an important factor in extending the outreach of glycomics is the urgent need for databases to store, process and disseminate structural and analytical datasets. The widely acknowledged sparseness of maintained resources continues to hamper the realisation tools, which are increasingly required to support high-throughput glycomics. During the 2nd Beilstein Glyco-Bioinformatics Symposium 2011 members of the UniCarbKB consortium presented a long-term vision [2] to build an infrastructure that adopts and extends the principles of quality shared by GlycoSuiteDB [3] and EUROCarbDB [4]. This new open-access infrastructure (called UniCarbKB) shall constitute the nucleus for a central depository for carbohydrate-related data (structure and function), comparable to and cross-linked with the extensively used genomics and proteomics data collections.

UniCarbKB: Laying the Foundations in Year One

UniCarbKB strives to radically enhance the infrastructure required to better enable glycomics research by making data more accessible, and presenting a single, user-friendly interface to a growing range of resources being developed. UniCarbKB was officially launched at the 3rd Beilstein Glyco-Bioinformatics Symposium 2013 [5]. The launch highlighted our mission to support glycomics and glycobiology research by providing a comprehensive, richly and accurately annotated glycan structure knowledgebase, with extensive cross-references and a redesigned intuitive user interface. The project is maturing and will continue to do so, but even at this early stage it has started to address major concerns raised by the research community and issues outlined in the 2012 NIH report.

For year one our design ethos has focused on providing researchers with 1) long-term and open-access to a highly-curated database of experimentally determined glycan structures; by 2) establishing a centralised model for merging curated data collections from GlycoSuiteDB, GlycoBase [6], UniCarb-DB [7] and EUROCarbDB; and 3) increasing awareness of data collections by connecting UniCarbKB with other databases such as SugarBind, a database of pathogen–sugar interactions [8]. Such activities include the refactoring of GlycoSuiteDB and the modification of modules from EUROCarbDB necessary to support existing and future structure and analytical data collections. During the engineering phase the developers systematically explored the functionality and database designs of EURO-CarbDB and GlycoSuiteDB. Here, the database structure integrates components from previous efforts to provide a more flexible and comprehensive relational database. It is likely that the schema will continually evolve with new data collections and data requirements. For example, core features of EUROCarbDB have been retained including GlycanBuilder [9, 10], MonosaccharideDB (http://www.monosaccharidedb.org) and the encoding standards developed in the context of GlycoCT [11].

Early phase developments are focused on enhancing existing tools, standards and applications to be more accessible and amenable to modern research workflows. In particular we have leveraged previous experiences to build a modern and scalable framework, which uses technologies and web frameworks that are more familiar to developers.

Establishment of the eResearch Infrastructure

Support from the Australian National eResearch Collaboration Tools and Resources (NeCTAR) provides UniCarbKB and its affiliated activities access to a sustainable infrastructure. A core principle of NeCTAR is to manage and preserve valuable data collections through the provision of robust data service architecture. More importantly the significant data centric investments will be maintained and shall form a critical component of life science research, thus ensuring that data will not disappear. In addition, the longevity of the resource shall be optimised by the mirroring of UniCarbKB on the ExPASy server [12], and by establishing close ties with the UniProt protein knowledgebase [13]. In the initial scope of the project the partners are establishing a framework of connected services comprising 1) a dedicated web-application front-end hosting the searchable main glycan structural database and 2) a mass spectrometry data collection supported by the efforts of UniCarb-DB and 3) the development of processing and interpretation tools.

Designing the UniCarbKB Experience

We have worked hard to put user experience design at the heart of the development cycle. When starting UniCarbKB the developers interacted with researchers and identified a series of 'stories' or specific themes that would need to be supported. It was critical, at an early stage, to identify the requirements and expectations of the community; for example we asked what features of EUROCarbDB and GlycoSuiteDB users frequently used and if/what improvements need to be considered. A series of design sprints and wire-framing sessions were conducted between the developers and the identified user base. Design sprints included a series of prototypes and iterative usability-based refinements of the design, evaluation of page layout, and generation of wireframes. By adopting agile development practices the developers and designers were able to build an application that is easy to use and allows researchers to accomplish their goals. This is being continually tested by a core group of users for feedback and improvements.

A detailed description of the new user-interface is documented in the *Nucleic Acids Research Database Issue* [5]. In brief, the new interface is more visual, encapsulating a simpler content layout with an emphasis placed on displaying information that researchers want to access. Many of these changes are in line with the presentation of information previously available from EUROCarbDB and GlycoSuiteDB databases and a choice of symbol nomenclatures [14, 15]. The goal is to retain the established user-base and minimise distraction with the launch of UniCarbKB. The web-application user experience is built using Twitter Bootstrap that provides designers access to rich features namely: responsive design, Javascript interaction, typography and cross-browser compatibility. It provides a faster, easier and less repetitive solution for delivering a robust front-end. By making use of these features plus the integration of JavaScript libraries (notably jQuery) UniCarbKB is delivering an enhanced user experience.

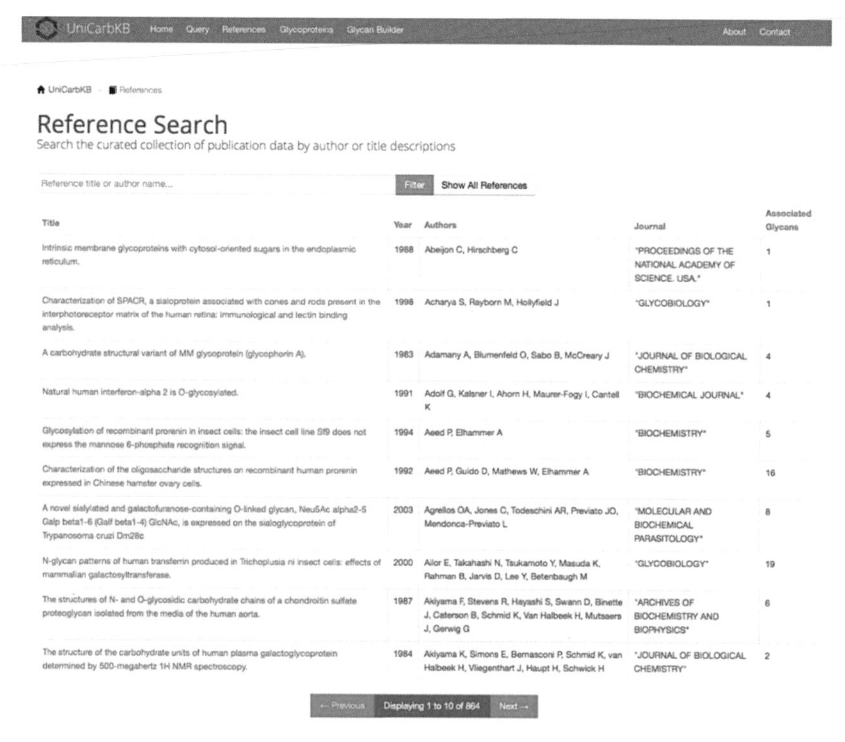

Title	Year	Authors	Journal	Associated Glycans
Intrinsic membrane glycoproteins with cytosol-oriented sugars in the endoplasmic reticulum.	1988	Abeijon C, Hirschberg C	"PROCEEDINGS OF THE NATIONAL ACADEMY OF SCIENCE. USA."	1
Characterization of SPACR, a sialoprotein associated with cones and rods present in the interphotoreceptor matrix of the human retina: immunological and lectin binding analysis.	1998	Acharya S, Rayborn M, Hollyfield J	"GLYCOBIOLOGY"	1
A carbohydrate structural variant of MM glycoprotein (glycophorin A).	1983	Adamany A, Blumenfeld O, Sabo B, McCreary J	"JOURNAL OF BIOLOGICAL CHEMISTRY"	4
Natural human interferon-alpha 2 is O-glycosylated.	1991	Adolf G, Kalsner I, Ahorn H, Maurer-Fogy I, Cantell K	"BIOCHEMICAL JOURNAL"	4
Glycosylation of recombinant prorenin in insect cells: the insect cell line Sf9 does not express the mannose 6-phosphate recognition signal.	1994	Aeed P, Elhammer A	"BIOCHEMISTRY"	5
Characterization of the oligosaccharide structures on recombinant human prorenin expressed in Chinese hamster ovary cells.	1992	Aeed P, Guido D, Mathews W, Elhammer A	"BIOCHEMISTRY"	16
A novel sialylated and galactofuranose-containing O-linked glycan, Neu5Ac alpha2-5 Galp beta1-6 (Galf beta1-4) GlcNAc, is expressed on the sialoglycoprotein of Trypanosoma cruzi Dm28c	2003	Agrellos OA, Jones C, Todeschini AR, Previato JO, Mendonca-Previato L	"MOLECULAR AND BIOCHEMICAL PARASITOLOGY"	8
N-glycan patterns of human transferrin produced in Trichoplusia ni insect cells: effects of mammalian galactosyltransferase.	2000	Ailor E, Takahashi N, Tsukamoto Y, Masuda K, Rahman B, Jarvis D, Lee Y, Betenbaugh M	"GLYCOBIOLOGY"	19
The structures of N- and O-glycosidic carbohydrate chains of a chondroitin sulfate proteoglycan isolated from the media of the human aorta.	1987	Akiyama F, Stevens R, Hayashi S, Swann D, Binette J, Caterson B, Schmid K, Van Halbeek H, Mutsaers J, Gerwig G	"ARCHIVES OF BIOCHEMISTRY AND BIOPHYSICS"	6
The structure of the carbohydrate units of human plasma galactoglycoprotein determined by 500-megahertz 1H NMR spectroscopy.	1984	Akiyama K, Simons E, Bernasconi P, Schmid K, van Halbeek H, Vliegenthart J, Haupt H, Schwick H	"JOURNAL OF BIOLOGICAL CHEMISTRY"	2

Figure 1. UniCarbKB has a completely revamped user-interface that focuses on presenting information in a clean and concise manner. For example, when browsing the reference collections users can search for content using the improved 'Filter' bar; searches can include author name or publication keywords. Pagination is used throughout UniCarbKB to improve both data layout and content navigation. The design of UniCarbKB was achieved by actively working alongside researchers to provide an intuitive data rich experience.

For example, to simplify searching and to organise result outputs we have made better use of auto completion, pagination and managed the efficient display of associated data links. Figure 1 shows how the user can readily search the curated publication information and pagination allows for better handling of large result sets. Figure 2 shows the summary page for the protein alpha-1-acid glycoprotein, which provides a description of the attached glycan structures and knowledge of site-specific glycosylation that has been curated from the literature. In addition, for each protein summary we provide a comprehensive summary of associated metadata including biological source and publications citing the data.

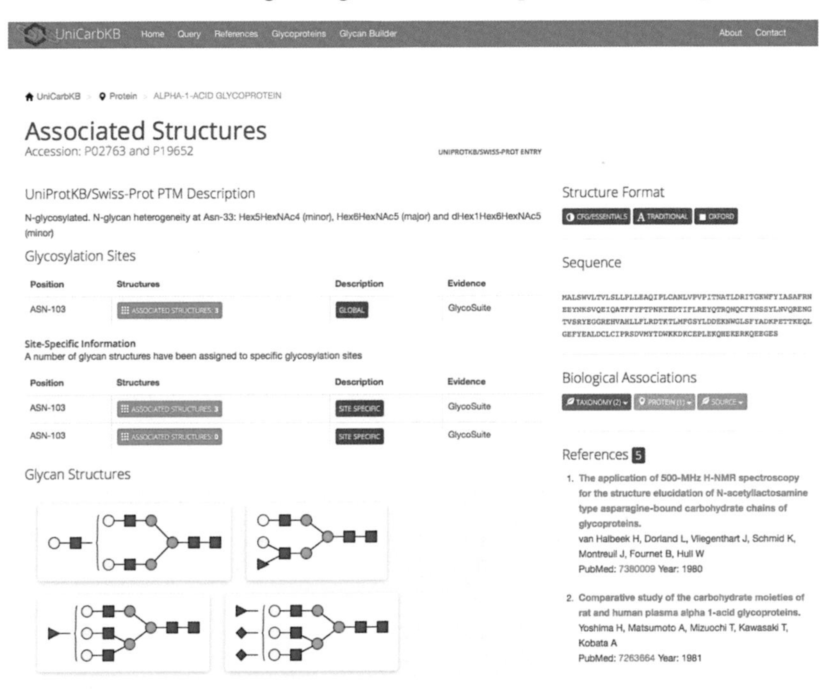

Figure 2. An example protein summary page for alpha-1-acid glycoprotein. The entry provides a description of the glycan structures characterised on this glycoprotein and the number of structures associated with experimentally confirmed glycosylation sites. Information is provided on the 'Biological Associations' including details on species and tissue source in addition to the protein sequence extracted from UniProtKB. The 'References' that have been used to curate the information are summarised in the right side panel including PubMed links.

UniCarbKB Biocuration

At this stage in the project UniCarbKB's central focus is the manual annotation of structural information from the literature, limited to mammalian *N*- and *O*-linked glycan structures. Journal articles provide the main source of experimental knowledge, with the full text of each article being read and the information on glycan composition, sequence, linkage, tissue and species source, attached protein if known, and the methods used to determine these characteristics being extracted manually. The aim of this approach is to build a central hub for glycan structures with an emphasis on quality glycoprotein information at the global and site-specific level. This has the advantage of providing a gold standard set of structural data in the knowledgebase, which can be used to build pipelines for the automatic capture of related information.

With the improvements in mass spectrometric technology, UniCarbKB is expanding from the initial focus of GlycoSuiteDB and EUROCarbDB in only collecting global glycoprotein glycan structures as described in the literature to now include site-specific glycosylation when known. For example, UniCarbKB curators are mining a literature review of 117 research papers encompassing over 400 glycosylation sites from more than 160 mammalian *N*-glycoproteins [16]. On-going efforts by the small curation team to include these data collections in the UniCarbKB knowledgebase will add considerable value to the existing 400 glycoproteins and 598 glycosylation sites now available. This site-specific glycosylation information will be linked to the relevant glycoprotein entries in UniProtKB.

Web Services and Semantic Technologies

Data-sharing and open-access models are increasingly important in the dissemination of knowledge in the life sciences. Increasingly, developers and users of bioinformatic platforms require access to well-documented libraries to create applications that query and make use of data collections available in public resources. To this end UniCarbKB supports programmatic access via web services to further facilitate the reuse of curated data collections across tools, other databases and research disciplines.

The web services build upon the efforts initiated by the Working Group for Glycomics Data(base) Standards (WGGDS) supported by the Consortium of Functional Glycomics (CFG) in 2010. In brief, the WGGDS was tasked with designing a cross-database interface based on RESTful protocols for querying and retrieving content from affiliated databases. To support the philosophy of UniCarbKB by providing the community with access to a high-quality database infrastructure the team have reengineered and extended components of the original protocols. UniCarbKB-WS is being developed with the JBossWS framework using enterprise standard protocols to provide the community with a first-of-its-kind secure and readily accessible remote querying resource.

A beta-release of UniCarbKB-WS is in the final stage of user testing prior to public release. Access to this resource, hosted on Australian NeCTAR infrastructure and mirrored on the Swiss ExPASY servers, will allow users to query the database via (sub)structure searching, structure IDs and glycan mass. A detailed description of the services available to developers is provided at http://dev.unicarbkb.org with an envisaged launch in early 2014.

Although the web services and query features on the website are efficient search tools there is an increasing demand in the life sciences for the provision of semantic web technologies. This is exemplified by the launch of the EMBL-EBI RDF platform (October 2013) that adopts the SPARQL Protocol and RDF Query Language, which provides a unified mechanism to query across multiple resources. In order to better integrate glycomics data collections, a glycoRDF working group comprising glycodatabase developers from Japan, Australia, Russia, Germany and USA has been established [17] and a standardised RDF document that is representative of all major glycan databases is being formalised. The access to a unified model will forge connections between existing resources that will considerably enhance data discovery and allow researchers to ask more complex biological questions. To this end, UniCarbKB developers are now contributing towards the generation of a compatible RDF framework, in agreement with members of the project, that includes (but is not limited to) a dedicated SPARQL endpoint and access to a regularly updated triplestore.

Connecting UniCarbKB with UniProtKB

There have been few intensive programmes to cross-reference glycan related databases with those that support genomics and proteomics research. Previously, GlycoSuiteDB was the sole provider of curated glycoprotein data to UniProtKB [12], however, during the last year members from both database activities have commenced a programme to share new glyco-protein knowledge through agreed formats. This collaborative effort, although in its infancy, aims to establish a mechanism whereby new content gained by UniCarbKB is integrated into UniProtKB on a regular basis.

A milestone towards this long-term agreement was reached in May 2013 with UniProtKB switching to UniCarbKB links instead of (the no longer maintained) GlycoSuiteDB (UniProt release 2013_06). Subsequently, in the UniProt release 2013_08 the GlycoSuiteDB name was substituted by UniCarbKB; such that all UniProtKB glycosylation entries associated with GlycoSuiteDB have now been updated to link with UniCarbKB and *vice-versa*.

To ensure accurate cross-referencing the developers standardised and defined ontology terms to link the two databases. Updates to relevant UniProtKB records and the inclusion of new unique identifiers (IDs) to UniCarbKB glycan structures allow users direct access to structural information and corresponding meta-data for relevant glycoproteins. In agreement with senior developers at UniProtKB new glycoprotein data curated by UniCarbKB will be integrated into UniProtKB and information flow will be two-way between the databases.

GlycoMod: Linking Monosaccharide Composition with Glycan Structures

GlycoMod [18] is a well-established programme designed to determine possible glycan structure compositions from experimentally determined mass. The tool is part of the proteomics suite of tools available on ExPASy. Indeed, it was an innovative tool that in 2001 recognised the importance of connecting proteomics and glycomics. The programme can be used to predict the composition of a glycan comprised of either underivatised, methylated or acetylated monosaccharides, or with a derivatised reducing terminus. Furthermore, the composition of a glycan attached to a glycopeptide can be calculated if the sequence or mass of the attached peptide is known. In such instances GlycoMod communicates with the UniProt Knowledgebase databases and matches experimentally determined glycopeptide masses against predicted protease-produced peptides, which enables the prediction of the composition of *N*- and *O*-linked oligosaccharides on glycopeptides.

We have directly linked the GlycoMod tool with UniCarbKB. Here, the two database initiatives actively share compositional and glycan structural data respectively. For each theoretical composition calculated by GlycoMod links to UniCarbKB will provide users with a list of composition-structure matches reported in the literature.

Integrating N-glycan Biosynthesis Pathways to Help Validate Structure Entries

Recently, we have initiated a project to integrate information about known genes and enzymes involved in the biosynthesis of *N*-glycans. This new platform, called GlycanSynth, involves the comprehensive curation of data related to enzyme activity, which is primarily sourced from the Encyclopedia of Genes and Genomes (KEGG) [19] and GlycoGene [20] databases. Additional information is also being gathered from the Consortium for Functional Glycomics (CFG) [21], Carbohydrate-Active enzymes (CAZy) [22, 23], BRENDA [24] and UniProt databases.

From this acquired information we have constructed a set of disaccharide reactions that match each *N*-glycosylation-related gene against donor and acceptor substrate. By using these reaction rules it will be possible for us to (i) connect gene function with glycan structure and (ii) validate the accuracy of structures stored in UniCarbKB based on acquired knowledge of the glycosylation machinery.

Forging New Connections with PubChem

PubChem (http://pubchem.ncbi.nlm.nih.gov) [25] is a public repository of molecules and their biological properties, containing more than 25 million chemical structures and 90 million bioactivity outcomes. It is the mission of PubChem to deliver free and easy access to deposited data, and to provide intuitive data analysis tools. The platform consists of three interconnected databases: Substance, Compound and BioAssay. It is the Substance database that contains detailed descriptions of molecules provided by depositors; the Compound database consists of unique chemical structures derived by structural standardisation of the records in the Substance section; while the BioAssay provides screening results of substances by assay providers.

In April 2013 PubChem and UniCarbKB announced the first phase of a programme to share glycan structural information. The aim of the programme is to increase the number and quality of biologically relevant protein-attached glycans available to researchers in PubChem. Stage one focused on validating PubChem's workflow for handling and processing a subset of fully defined UniCarbKB structures encoded in the IUPAC format. PubChem has been collaborating with NextMove on the development of 'Sugar & Splice' to convert representation of glycans. This was the first large activity that required the conversion of IUPAC notation to the SMILES representation. This collaboration will continue to grow with UniCarbKB providing regular updates to PubChem in parallel with future curation plans.

Conclusion

In 2011 we announced plans to implement a glycosciences knowledge platform that would set the state- of-the-art foundation infrastructure for the free sharing and dissemination of data on glycoconjugates, and which would seed the development of enhanced glyco-informatic tools with which to interpret the data. By 2013, in collaboration with international partners and with limited resources, we have made significant strides toward meeting our ambitious goals in a short time span.

Such efforts have spanned the development of a new web-application framework with an entirely new approach to user-interface designs and the introduction of new technologies including RDF Semantics. Considerable advancements have afforded 1) an amalgamated glycan structure database that brings together collections curated and sourced from legacy efforts; 2) a technical framework that is now enabling the consortia to deploy tools and workflows to assist the interpretation of experimental (principally mass spectrometry at this stage) data; 3) a biocuration programme to capture newly published data to significantly extend the availability of high-quality datasets and 4) established crosslinks with the proteomics knowledgebase UniProtKB and the chemical repository PubChem. Our biocuration efforts have also provided a forum for us to open discussions with Thomson

Reuters Data Citation Index (DCI) to adopt UniCarbKB as the prime indexing site for glycomics. This activity stems from our work with the Australian National Data Services (ANDS) to promote the discovery and re-use of research data. It is envisaged that early in year two DCI-ANDS will harvest data directly from UniCarbKB to link the resources directly; offering users an enhanced data search platform.

UniCarbKB is a user driven resource and feedback is extremely valuable to help us improve the content of our databases and the services offered in terms of accuracy and usability. To promote active outreach we have released a feedback tool (accessible on all pages of UniCarbKB) for users to provide comments. We are keen to work with researchers to include newly published data or updates, and as such we strongly encourage the use the forms available at http://www.unicarbkb.org/contribute to engage with our small data curation team. New releases will be published every three months accompanied by a description of changes and inclusions.

UniCarbKB has been acknowledged as a necessary, valuable and quality resource for the glycoscience community. Strong foundations for the expansion of the knowledgebase have now been set but require international buy-in with real resources for its continued development.

Acknowledgements

The authors are grateful for funding support from: The Australian National eResearch Collaboration Tools and Resources project [NeCTAR RT016 to M.P.C and N.H.P]; Australian National Data Service (ANDS) through the National Collaborative Research Infrastructure Strategy Program and the Education Investment Fund (EIF) Super Science Initiative to M.P.C and N.H.P]; Swiss National Science Foundation [SNSF 31003A_141215 J.M.]; Swiss Federal Government through the State Secretariat for Education, Research and Innovation SERI [F.L. and E.G.]; ExPASy is maintained by the web team of the Swiss Institute of Bioinformatics and hosted at the Vital-IT Competency Center; UniCarbKB was also supported by Agilent's University Relations programme to M.P.C and N.H.P; GlycoSuiteDB was developed by Proteome Systems Ltd [N.H.P] and transferred to SIB in 2009; The Swedish Foundation for International Cooperation in Research and Higher Education to M.P.C, C.A.H and N.G.K.

References

[1] Transforming Glycoscience: A Roadmap for the Future. 2012, National Research Council.: Washington (DC).

[2] Campbell, M.P., Hayes, C.A., Struwe, W.B., Wilkins, M.R., Aoki-Kinoshita, K.F., Harvey, D.J., Rudd, P.M., Kolarich, D., Lisacek, F., Karlsson, N.G., Packer, N.H. (2011) UniCarbKB: Putting the pieces together for glycomics research. *Proteomics* **11**:4117 – 4121.
doi.org/ 10.1002/pmic.201100302.

[3] Cooper, C.A., Harrison, M.J., Wilkins, M.R., Packer, N.H. (2001) GlycoSuiteDB: a new curated relational database of glycoprotein glycan structures and their biological sources. *Nucleic Acids Research* **29**:332 – 335.
doi: 10.1093/nar/29.1.332.

[4] von der Lieth, C.-W., Freire, A.A., Blank, D., Campbell, M.P., Ceroni, A., Damerell, D.R., Dell, A., Dwek, R.A., Ernst, B., Fogh, R., Frank, M., Geyer, H., Geyer, R., Harrison, M.J., Henrick, K., Herget, S., Hull, W E., Ionides, J., Joshi, H.J., Kamerling, J.P., Leeflang, B.R., Lütteke, T., Lundborg, M., Maass, K., Merry, A., Ranzinger, R., Rosen, J., Royle, L., Rudd, P.M., Schloissnig, S., Stenutz, R., Vranken, W.F., Widmalm, G., Haslam, S.M. (2011) EUROCarbDB: An open-access platform for glycoinformatics. *Glycobiology* **21**:493 – 502.
doi: 10.1093/glycob/cwq188.

[5] Campbell, M.P., Peterson, R., Mariethoz, J., Gasteiger, E., Akune, Y., Aoki-Kinoshita, K. F., Lisacek, F., Packer, N.H. (2013) UniCarbKB: building a knowledge platform for glycoproteomics. *Nucleic Acids Research* **42**:D215 –D221.
doi: 10.1093/nar/gkt1128.

[6] Campbell, M.P., Royle, L., Radcliffe, C.M., Dwek, R.A., Rudd, P.M. (2008) GlycoBase and autoGU: tools for HPLC-based glycan analysis *Bioinformatics* **24**(9):1214 – 1216.
doi: 10.1093/bioinformatics/btn090.

[7] Hayes, C.A., Karlsson, N.G., Struwe, W.B., Lisacek, F., Rudd, P.M., Packer, N.H., Campbell, M.P. (2011) UniCarb-DB: a database resource for glycomic discovery. *Bioinformatics* **27**:1343 – 1344.
doi: 10.1093/bioinformatics/btr137.

[8] Shakhsheer, B., Anderson, M., Khatib, K., Tadoori, L., Joshi, L., Lisacek, F., Hirschmann, L, Mullen, E. (2013) SugarBind database (SugarBindDB): a resource of pathogen lectins and corresponding glycan targets. *Journal of Molecular Recognition* **26**(9):426 – 431.
doi: 10.1002/jmr.2285.

[9] Ceroni, A., Dell, A., Haslam, S.M. (2007) The GlycanBuilder: a fast, intuitive and flexible software tool for building and displaying glycan structures. *Source Code for Biology and Medicine* **2**:3.
doi: 10.1186/1751-0473-2-3.

[10] Damerell, D., Ceroni, A., Maass, K., Ranzinger, R., Dell, A., Haslam, S.M. (2012) The GlycanBuilder and GlycoWorkbench glycoinformatics tools: updates and new developments. *Biological Chemistry* **393**(11):1357 – 1362.
doi: 10.1515/hsz-2012-0135.

[11] Herget, S., Ranzinger, R., Maass, K., von der Lieth, C.-W. (2008) GlycoCT – a unifying sequence format for carbohydrates. *Carbohydrate Research* **343**(12): 2162 – 2171.
doi: 10.1016/j.carres.2008.03.011.

[12] Artimo, P., Jonnalagedda, M., Arnold, K., Baratin, D., Csardi, G., de Castro, E., Duvaud, S., Flegel, V., Fortier, A., Gasteiger, E., Grosdidier, A., Hernandez, C., Ioannidis, V., Kuznetsov, D., Liechti, E., Moretti, S., Mostaguir, K., Redaschi, N., Grégoire Rossier, G., Xenarios, I., Stockinger, H. (2012) ExPASy: SIB bioinformatics resource portal. *Nucleic Acids Research* **40**(Web Server issue): W597 –W603.
doi: 10.1093/nar/gks400.

[13] The UniProt Consortium (2013) Update on activities at the Universal Protein Resource (UniProt) in 2013. *Nucleic Acids Research* 2013. **41**(Database issue): D43 –D47.
doi: 10.1093/nar/gks1068.

[14] Varki, A., Cummings, R.D., Esko, J.D., Hudson H. Freeze, H.H., Stanley, P., Marth, J.D., Bertozzi, C.R., Hart, G.W., Etzler, M.E. (2009) Symbol nomenclature for glycan representation. *Proteomics* **9**(24):5398 – 5399.
doi: 10.1002/pmic.200900708.

[15] Harvey, D.J., Merry, A.H., Royle, L., Campbell, M.P., Dwek, R.A., Rudd, P.M. (2009) Proposal for a standard system for drawing structural diagrams of *N*- and *O*-linked carbohydrates and related compounds. *Proteomics* **9**(15):3796 – 3801.
doi: 10.1002/pmic.200900096.

[16] Thaysen-Andersen, M., Packer, N.H. (2012) Site-specific glycoproteomics confirms that protein structure dictates formation of *N*-glycan type, core fucosylation and branching. *Glycobiology* **22**(11):1440 – 1452.
doi: 10.1093/glycob/cws110.

[17] Aoki-Kinoshita, K.F., Bolleman, J., Campbell, M.P., Kawano, S., Kim, J.-D., Lütteke, T., Matsubara, M., Okuda, S., Ranzinger, R., Sawaki, H., Shikanai, T., Shinmachi, D., Suzuki, Y., Toukach, P., Yamada, I., Packer, N.H., Narimatsu, H. (2013) Introducing glycomics data into the Semantic Web. *Journal of Biomedical Semantics* **4**(1):39.
doi: 10.1186/2041-1480-4-39.

[18] Cooper, C.A., Gasteiger, E., Packer, N.H. (2001) GlycoMod – a software tool for determining glycosylation compositions from mass spectrometric data. *Proteomics* **1**(2):340 – 349.
doi: 10.1002/1615-9861(200102)1:2<340::AID-PROT340>3.0.CO;2-B.

[19] Kanehisa, M., Goto, S. (2000) KEGG: kyoto encyclopedia of genes and genomes. *Nucleic Acids Research* **28**(1):27 – 30.
doi: 10.1093/nar/28.1.27.

[20] Narimatsu, H. (2004) Construction of a human glycogene library and comprehensive functional analysis. *Glycoconjugate Journal* **21**(1 – 2):17 – 24.
doi: 10.1023/B:GLYC.0000043742.99482.01.

[21] Raman, R., Venkataraman, M., Ramakrishnan, S., Lang, W., Raguram, S., Sasisekharan, R. (2006) Advancing glycomics: implementation strategies at the consortium for functional glycomics. *Glycobiology* **16**(5):82R– 90R.
doi: 10.1093/glycob/cwj080.

[22] Lombard, V., Ramulu, H.G., Drula, E., Coutinho, P.M., Henrissat, B. (2013) The carbohydrate-active enzymes database (CAZy) in 2013. *Nucleic Acids Research* **42**(D 1):D 490 –D 495.
doi: 10.1093/nar/gkt1178.

[23] Cantarel, B.L., Coutinho, P.M., Rancurel, C., Bernard, T., Lombard, V., Henrissat, B. (2009) The Carbohydrate-Active EnZymes database (CAZy): an expert resource for Glycogenomics. *Nucleic Acids Research* **37**(Database issue):D 233 –D 238.
doi: 10.1093/nar/gkn663.

[24] Schomburg, I., Chang, A., Placzek, S., Söhngen, C., Rother, M., Lang, M., Munaretto, C., Ulas, S., Stelzer, M., Grote, A., Scheer, M., Schomburg, D. (2013) BRENDA in 2013: integrated reactions, kinetic data, enzyme function data, improved disease classification: new options and contents in BRENDA. *Nucleic Acids Research* **41**(Database issue):D 764 –D 772.
doi: 10.1093/nar/gks1049.

[25] Bolton, E.E., Wang, Y., Thiessen, P.A., Bryant, S.H. (2018) PubChem: Integrated Platform of Small Molecules and Biological Activities. *Annual Reports in Computational Chemistry* **4**:217 – 241.
doi: 10.1016/S1574-1400(08)00012-1

High-throughput Workflow for Glycan Profiling and Characterisation

Henning Stöckmann, Giorgio Carta, Ciara A. McManus, Mark Hilliard, and Pauline M. Rudd*

NIBRT GlycoScience Group, NIBRT – The National Institute for Bioprocessing Research and Training, Foster's Avenue, Mount Merrion, Blackrock, Co. Dublin, Ireland

E-Mail: *pauline.rudd@nibrt.ie

Received: 20th January 2014 / Published: 22nd December 2014

Abstract

Over the last 40 years, the understanding of glycosylation changes in health and disease has evolved significantly and glycans are now regarded as excellent biomarker candidates because of their high sensitivity to pathological changes. However, the discovery of clinical glycobiomarkers has been slow, mainly as a consequence of the lack of high-throughput glycoanalytical workflows that allow rapid glyco-profiling of large clinical sample sets. To generate high-quality quantitative glycomics data in a high-throughput fashion, we have developed a robotised platform for rapid *N*-glycan sample preparation and glycan characterisation. The sample preparation workflow features a fully automated, rapid glycoprotein affinity purification followed by sequential protein denaturation and enzymatic glycan release on a multiwell ultrafiltration device, thus greatly streamlining all required biochemical manipulations. After glycan purification on solid-supported hydrazide, glycans are fluorescently labelled to allow accurate quantification by ultra-high pressure liquid chromatography (ultra HPLC or UPLC). Subsequent peak assignment can be carried out utilising GlycoBase, a bespoke chromatographic data system developed to aid the analysis of glycans performed using different chromatographic techniques (UPLC, HPLC, Reverse Phase-UPLC, Capillary Electrophoresis).

This article is part of the Proceedings of the Beilstein Glyco-Bioinformatics Symposium 2013.
www.proceedings.beilstein-symposia.org

INTRODUCTION

Complex carbohydrates occur in a variety of forms and locations throughout the body and it is estimated that over half of eukaryotic proteins are in fact glycosylated [1]. Modern biopharmaceuticals are glycosylated, which can have a profound impact on pharmacodynamics and pharmacokinetics. *N*- and *O*-linked glycans are attached to both cell surfaces and secreted proteins. In turn, the diversity of possible monosaccharide combinations and conformations gives rise to oligosaccharides and protein site attachment variations also increase heterogeneity. Both the number and the structural diversity of glycans found in mammalian systems present significant analytical challenges for determining detailed glycan structural profiles in complex organisms. Glycosylation plays fundamental roles in many biological recognition events. Glycans undergo rapid structural changes in response to biological stimuli, providing a unique opportunity to identify and exploit glycans as clinical markers that can be indicative of specific disease states, disease progression, and/or therapy response. The concept of biomedical glycomics has gained considerable momentum and efforts are underway to rapidly identify glycans as disease biomarkers, to reveal the mechanisms that regulate glycan biosynthetic pathways, and to decode the functions of glycans in complex biological systems. In particular, large-scale glycoprofiling is being conducted in combination with human genome-wide association studies (GWAS) to understand the complex regulation of glycan expression in humans (Figure 1).

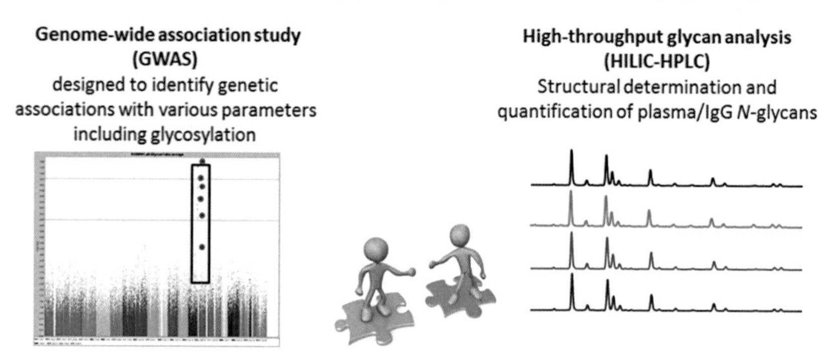

Figure 1. The 'Genomics meets Glycomics' concept. Genome-wide association studies (GWAS) are conducted to determine genetic variations that contribute to polygenic diseases, such as cancer and diabetes. Genetic associations can help develop improved strategies to detect, treat and prevent the disease. Glycome-screening studies help identify correlations between a disease phenotype and certain glycan features across glycomes. Linking GWAS and glycomics data unravels the regulation of glycan expression.

GWAS aim to identify genetic variations associated with particular diseases and involve the rapid screening of single-nucleotide polymorphism (SNP) markers across the genomes of thousands of individuals. When GWAS data are correlated with glycomics data, glycomic variations can be associated with genomic variations to elucidate the regulation of glycans, which can provide insights into pathophysiology. For example, linking GWAS data and glycomics data has been used to identify HNF1a as a master regulator of fucosylation [2].

Despite the great promise of GWAS-linked glycomics studies, low-cost technologies to rapidly obtain quantitative glycomics data from large sample numbers have not been developed. Although there is pressing need for high-throughput, reliable workflows in biomedical glycomics, biopharmaceutical development also require high-throughput glyco-profiling technologies, for example in the selection of clones that produce monoclonal antibodies or hormones with a particular glycosylation pattern.

High-Throughput Glycan Profiling Platform

We had previously developed a high performance liquid chromatography (HPLC)-based analysis for serum *N*-glycans in a 96-well format [3], and this has been successfully utilised in glycoprofiling studies in complex diseases such as cancer and diabetes [4 – 7]. However, while 'high-thoughput' can be carried out by intensive manual labour, the processing time can be up to three days for 96 samples and the method is difficult to automate. Thus, we have now developed the first fully automated, cost-effective multi-purpose glycomics platform, thereby considerably expanding the existing repertoire of glycomics workflows [8]. The main purpose of the platform is accurate glycan quantification in complex biological samples such as patient sera and cell culture supernatants. The platform is versatile and can be used to isolate individual glycoproteins or classes of glycoproteins (such as immunoglobulin G) and determine their glycosylation pattern. Glycoprotein samples are first prepared on a liquid handling robot (described in more detail below), and glycans are then separated with ultra-high performance liquid chromatography (UPLC), resulting in fluorescence chromatograms. Peak annotation is conducted using glycan sequencing and reference data stored in a dedicated database, termed GlycoBase (Figure 2). In addition, glycans can be labelled with different fluorophores, separated by a range of separation technologies both individual and coupled like LC/MS.

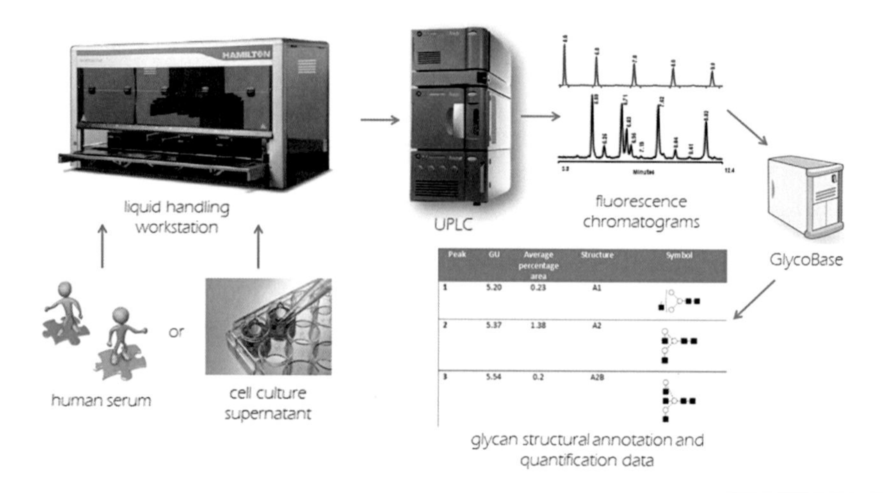

Figure 2. Glycomics workflow. Sera or cell culture supernatant samples are processed on a robotic workstation, resulting in fluorescently labelled glycans, which are subsequently separated and quantified by ultra-high pressure liquid chromatography (UPLC). Glycan peak annotations are performed based on enzymatic glycan sequencing and glucose unit (GU) data stored in the GlycoBase database.

Efficient and high-throughput glycan sample preparation is a key element of the glycomics platform. The workflow (Figure 3) is initiated by an automated affinity purification step of the glycoprotein of interest, e.g., immunoglobulin G (IgG). Subsequently, proteins are denatured and *N*-glycans released from the protein enzymatically by Peptide-*N*-Glycosidase F (PNGaseF). All biochemical operations are performed on an ultrafiltration plate to efficiently remove buffer and excess reagents and to separate the glycans from the protein.

Glyans are then captured on solid supported hydrazide to enable the removal of residual impurities which can interfere with fluorescent labelling. First, reducing-end glycans are reacted with solid-supported hydrazides to form hydrazones, thus, capturing glycans on the solid support. Contaminants such as excess reagents and buffer salts are removed by filtration. Next, glycans are released by acid catalysis in the presence of water. To equip glycans with stoichiometric amounts of fluorophore, they are labelled by reductive amination with 2-aminobenzamide (2-AB), a well-known and reliable reaction. Post-labelling clean-up is performed by solid-phase extraction.

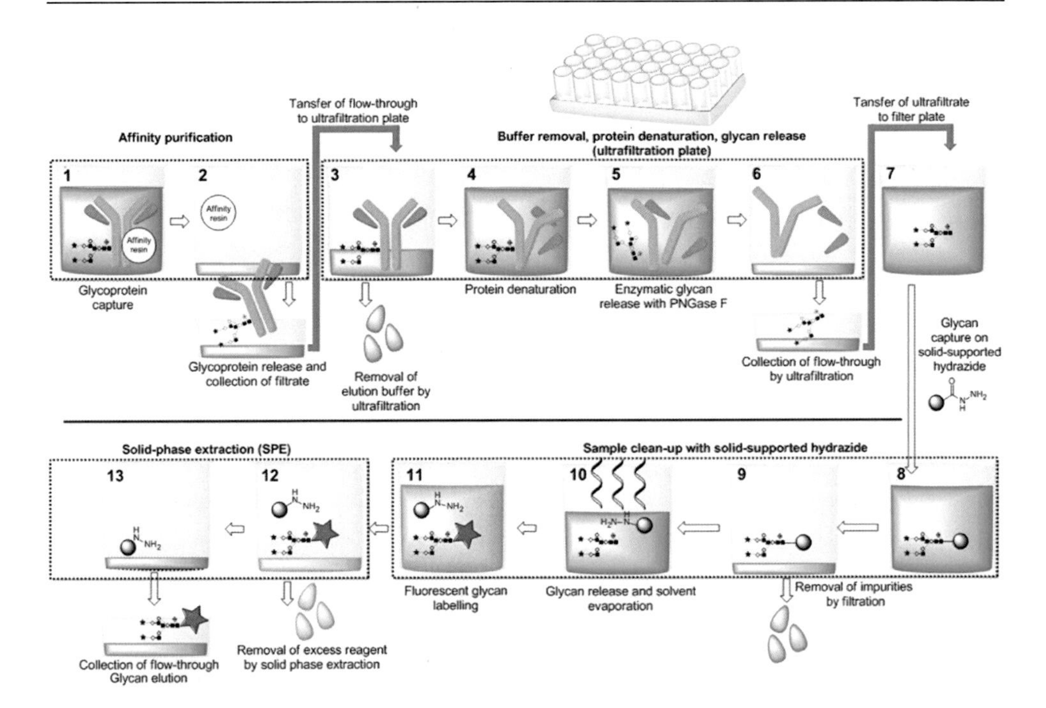

Figure 3. Sample preparation workflow on the NIBRT robotised platform. This schematic depicts the steps (1 – 13) for sample preparation, consisting of glycoprotein affinity purification (steps 1 – 2), buffer removal and protein denaturation (steps 3 – 4), enzymatic glycan release (step 5), glycan immobilisation on solid supports (step 8), removal of contaminants (step 9), glycan release (step 10), fluorescence labelling (step 11), and solid-phase extraction (steps 12 – 13). Reprinted and adapted with permission from reference [8]. Copyright 2013 American Chemical Society.

A robust robotic program with automatic error recovery was created to implement the workflow on a commercial liquid handling robot (Figure 4). The fully software-controlled workstation was equipped with eight robotic pipettes with individual liquid level and pressure sensors (Figure 4B), pipette tip racks, plate carriers, reagent reservoirs, a software-controlled vacuum manifold (Figure 4C), a temperature-controlled orbital shaker (Figure 4D), and a plate-transport tool to enable the movement of multi-well plates between positions and to operate the vacuum manifold.

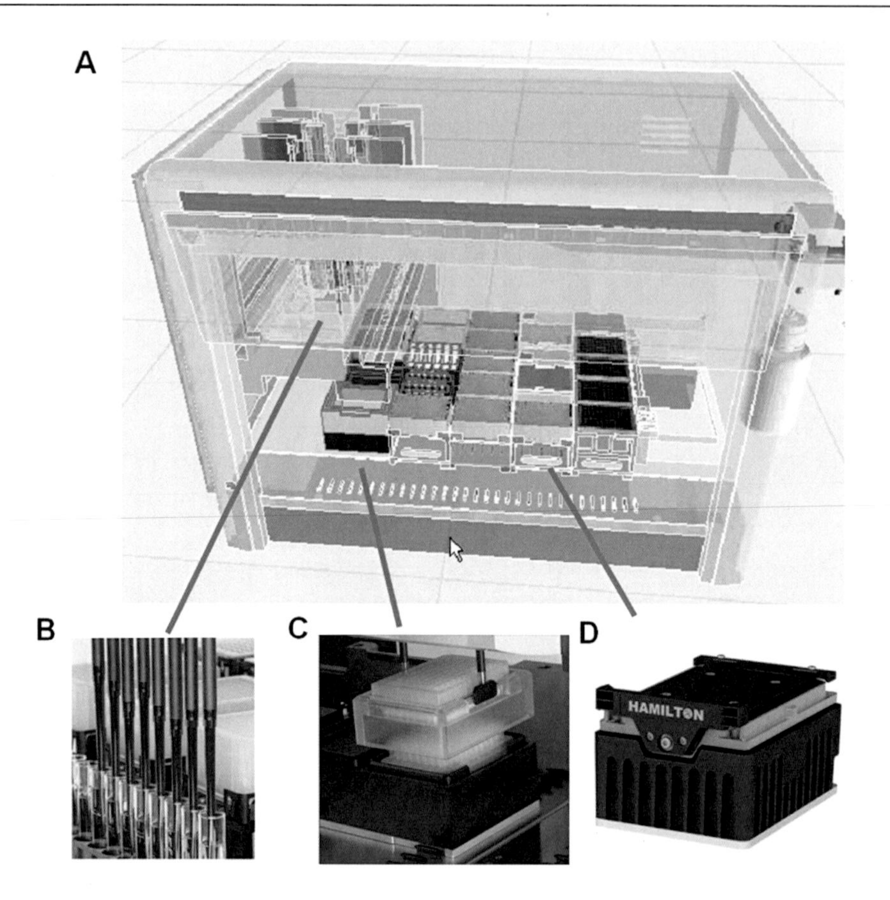

Figure 4. Schematic of the liquid-handling workstation utilised to perform high-throughput automated glycan sample preparation **(A)**. Key components include: pipetting channels with liquid-level detection and antidroplet control **(B)**, software-controlled robotic vacuum manifold and plate-transport tool **(C)**, and temperature-controlled orbital shaker **(D)**. Reprinted with permission from reference [8]. Copyright 2013 American Chemical Society.

The processing of up to 96 samples including glycoprotein affinity purification in a 96 well plate format typically takes around 22 h. The fluorescently labelled glycans are run on HPLC/UPLC instruments equipped with hydrophilic interaction chromatrography (HILIC) columns and the resulting peaks are correlated to a pre-run dextran ladder, thereby assigning a Glucose Unit (GU) value to each of the peaks. The use of standard glucose units makes these values independent of the running conditions; which allows for the direct comparison of chromatographic profile peaks and their relative glycan abundance (Figure 5). The coefficients of variation between samples prepared on different days with the automated robotised method for all major IgG peaks are typically below 10% (i.e., those peaks with a relative percentage area above 1%), indicating an excellent reproducibility.

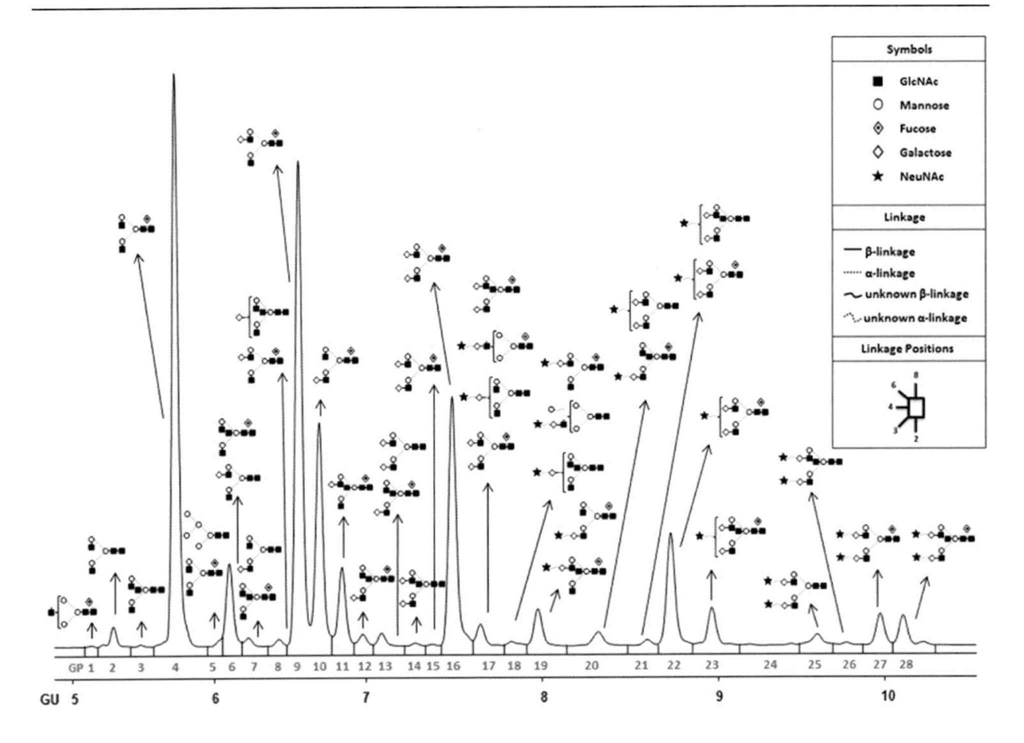

Figure 5. Human IgG N-glycome and peak assignments. IgG from native human serum was isolated and processed on the liquid-handling workstation followed by glycan analysis by ultra-high pressure liquid chromatography with fluorescence detection. GU: glucose units. Reprinted with permission from reference [8]. Copyright 2013 American Chemical Society.

Due to its robustness, high throughput and low cost, the platform is an ideal tool for efficient and accurate glycan profiling for GWAS and biopharmaceutical development and has been extensively used in these contexts (publications in preparation).

Glycan Structure Elucidation Using Glycan Sequencing and NIBRT's GlycoBase

Structure elucidation of glycan peaks requires reliable techniques and glycan reference data. Challenges in structural analysis include the large number of glycan classes and the efficient exploitation of analytical and bioinformatics tools that are available for structural interrogation. Data sources for glycan analytics encompass several orthogonal methodologies such as ultra-high pressure liquid chromatography (UPLC), capillary electrophoresis (CE) and mass spectrometry (MS), all of which have inherent difficulties in data interpretation. Assignment and characterisation of glycan structures in biotherapeutic products or high-throughput data from clinical profiling is a difficult and time-consuming process and is often a bottleneck in this type of research. It therefore requires automated data-integration, data-mining and statistical analysis tools coupled with software engineering and database technology to advance this field of research. This would bring glycomic analysis in line with both the proteomic and genomic fields.

NIBRT's GlycoBase (www.glycobase.nibrt.ie, [9]) is an integrated solution for rapid and reproducible characterisation of glycan samples. Originally developed from the database EurocarbDB, GlycoBase is a resource for the storage, classification and reporting of glycan structures as well as their associated experimental values obtained using various chromatographic techniques such as HPLC, UPLC and CE. GlycoBase is a web-enabled, open-access resource that contains glycan data as normalised chromatographic retention time data, expressed as GU values, for more than 740 2-AB labelled *N*-linked glycan structures. These values were experimentally obtained by systematic analysis of released *N*-glycans from a diverse set of glycoproteins on the NIBRT glycan analytical platform utilising both Waters HPLC and UPLC analytical instruments. The database was built using data from many samples over the course of one decade. The UPLC data were obtained from many analytes including human serum. The Waters collection is a list of GU values pertaining to the analyses of a number of therapeutically interesting glycoproteins including erythropoietin and herceptin, haptoglobin, RNAse B and transferrin and is continually being expanded. Hydrophilic interaction liquid chromatography combined with fluorescence detection (HILIC-fluorescence), supplemented by exoglycosidase sequencing and mass spectrometric confirmation, was used to generate this high confidence glycan library. The resulting database has been made accessible through a customised web-application containing a simple and intuitive interface to assign and confirm glycan structures.

GlycoBase enables users to search for specific glycans using a variety of searching tools. These include searching by the regular expression name or by antennary composition (e. g., A1, A2 etc.). Alternatively searches can be carried out according to a GU value (± 0.3), or the user can search for a particular glycan feature, for example the presence or absence of sialic acid or core-fucose. The user also has the ability to carry out a stoichiometric search and thus search by, for example the number of hexoses or xyloses. All the searches can be performed on a global basis, thus searching the entire collection, a selected collection or a particular sample within a collection. GlycoBase provides users with access to a "summary report" which collates all the available data for a selected glycan. This includes information on general glycan properties such as the monoisotopic mass and the monosaccharide composition. Individual experimental records containing for example all the UPLC derived GU values recorded in the database are also shown on this summary page. Similarly, the user can view links to literature records, profile information as well as the instrument running conditions.

Reliable glycan peak assignments and structure elucidation are achieved through GU data from GlycoBase combined with glycan sequencing. Glycan sequencing is performed by exoglycosidase glycan digestion and is an ideal method for rapid oligosaccharide characterisation including monosaccharide sequence and linkage information. Exoglycosidases remove carbohydrate residues from the non-reducing end of a glycan in a linkage-specific manner. For example, almond meal fucosidase (AMF) removes terminal α-fucose residues attached with a (1→3) or (1→4) linkage but not residues attached with a (1→6) linkage. In glycan sequencing, the glycan pool is analysed before and after sequential digestion with arrays of linkage-specific exoglycosidases. Glycan digestion results in peak shifts, the extent of which depends on the nature and the number of monosaccharides removed. The entire pool of glycans can be digested without separating individual peaks and aliquots of the pool can be digested simultaneously with panels of enzyme arrays. Figure 6 shows an example of a complete exoglycosidase digestion scheme for the structural analysis of a glycan pool obtained from Trastuzumab (trade name Herceptin), a monoclonal antibody used to treat certain types of breast cancer. Treatment of the glycan pool (i) with sialidase leads to the disappearance of two peaks at GU = 9.10 and GU = 8.33 and to a corresponding increase in the peak at GU = 7.60 (ii). A GU-shift of ca. 0.75 is characteristic for a sialic acid, so that the peaks at GU = 9.10 and GU = 8.33 must be glycans with two and one terminal sialic acids, respectively. The glycan pool obtained after sialidase digestion is then sequentially digested with fucosidase (iv), galactosidase (v) and hexosaminidase (vi), resulting in one single peak at GU = 4.30, which represents Man3GlcNAc2, the core structure of all *N*-linked glycans.

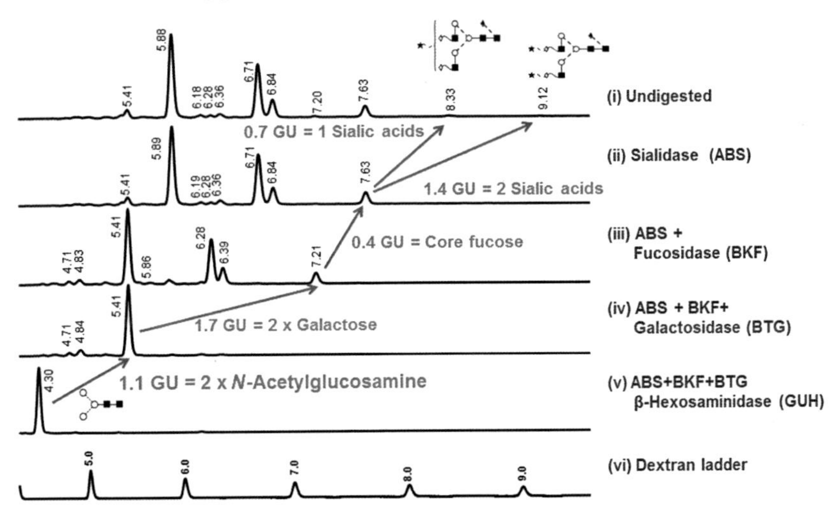

Figure 6. UPLC analysis and exoglycosidase array digestions of Trastuzumab (Herceptin) glycans analysed by UPLC with fluorescence detection. (i) Undigested glycan sample; (ii) ABS (*Athrobacter ureafaciens* sialidase) releases α(2 – 3,6,8)-linked sialic acids; (iii) ABS + BKF (Fucosidase from bovine kidney) releases α(1 – 2,6) linked fucose; (iv) ABS + BKF + BTG (Bovine testes β-galactosidase) releases β1 – 3 and 1 – 4 linkages, galactose and (v) ABS + BKF + BTG + GUH (hexosaminidase) releases β-GlcNAc but not GlcNAc linked to β(1 – 4) Man.

GlycoBase also enables the user to carry out *in silico* "GlycoBaseDigests". This predictive tool enables the generation of *in silico* exoglycosidase digests using various enzymes that are frequently used in the full characterisation of glycans. The user can select a particular glycan and perform a "virtual" digestion. GlycoBase returns the predicted digest product, and if experimental GU values associated with the queried and digested exist in the database, then these are also reported. GlycoBaseDigest currently provides *in silico* digestion for the following exoglycosidases: JBM, GUH, ABS, BKF, NAN1, AMF and SPG. (Figure 7).

Figure 7. Example of an *in silico* GlycoBase digest: The glycan A1G1 (GU = 5.7) is digested with SPG which releases β(1 – 4)-Galactosidase resulting in the glycan, A1 (GU = 4.8).

Additionally, GlycoBase allows users to conduct *in silico* "Extrapolated Profiling". GlycoBase stores glycan profiles using both the GU values as well as the area under the peak. Using this information and data, GlycoBase will then re-construct the original profile computing the Gaussian kernel density estimation. The relative percentage areas are used as a numeric vector of non-negative observation weights, while the GU values are the data from which the estimate is to be computed. The result is an *in silico* approximate profile (Figure 8). The computation is performed on the server side using the "R" statistical and graphic package.

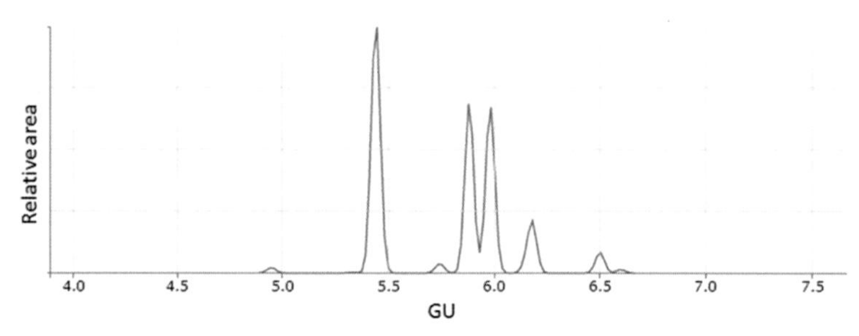

Figure 8. Example of an computationally calculated profile from human serum digested with a α(2 – 3,6,8) sialidase and a α(1 – 2,6) fucosidase.

If information regarding the exoglycosidase digest panels is available, GlycoBase users can then navigate from parent to child profiles. Thus, this allows the user to reconstruct the logic that was followed during the full characterisation of an undigested profile and enables the user to fully comprehend how each and every structure was identified.

Conclusion

The combined efficiencies of sample preparation, high resolution glycan analysis and computer assisted data processing can enable both the research community and the biopharmaceutical sector to perform released glycan analysis with greater confidence and speed than previously possible with the existing analysis workflows.

The robotic platform in combination with UPLC and GlycoBase as a robust database system, offer the basis for high-throughput glycoprofiling and characterisation of biological samples from biomarker discovery studies and clinical studies such as GWAS. While the robotic platform is an enabling technology to reduce processing time and cost associated with sample preparation, the UPLC method offers excellent peak resolution and performance for glycan analysis. The glycan sample preparation platform can be easily adapted and allows glycan labelling with a variety of labels so that it can be linked to complementary analytical technologies such as mass spectrometry and capillary electrophoresis.

Acknowledgments

The authors wish to acknowledge the following; GlycoBaseDigest: in collaboration with the Swiss Institute of Bioinformatics. GlycoBioM grant funding from the European Union Seventh Framework Programme FP7/2007 – 2013), Grant No 259869, HighGlycan grant funding from the European Union Seventh Framework Programme (FP7/2011 – 2013), Grant No 278535, HighGlycan

References

[1] Apweiler, R., Hermjakob, H., Sharon, N. (1999) On the frequency of protein glycosylation, as deduced from analysis of the SWISS-PROT database. *Biochimica et Biophysica Acta – General Subjects* **1473**:4 – 8. doi: 10.1016/S0304-4165(99)00165-8.

[2] Lauc, G., Essafi, A., Huffman, J. E., Hayward, C., Kneević, A., Kattla, J.J., Kattla, J.J., Polašek, O., Gornik, O., Vitart, V., Abrahams, J.L., Pučić, M., Novokmet, M., [...], Rudan, I. (2010) Genomics Meets Glycomics – The First GWAS Study of

Human *N*-Glycome Identifies HNF1α as a Master Regulator of Plasma Protein Fucosylation. *PLoS Genetics* **6**(12):e1001256.
doi: 10.1371/journal.pgen.1001256.

[3] Royle, L., Campbell, M.P., Radcliffe, C.M., White, D.M., Harvey, D.J., Abrahams, J.L., Kim, Y.-G., Henry, G.W., Shadick, N.A., Weinblatt, M.E., Lee, D.M., Rudd, P.M., Dwek, R.A. (2008) HPLC-based analysis of serum *N*-glycans on a 96-well plate platform with dedicated database software. *Analytical Biochemistry* **376**(1):1 – 12.
doi: 10.1016/j.ab.2007.12.012.

[4] Saldova, R., Royle, L., Radcliffe, C.M., Abd Hamid, U.M., Evans, R., Arnold, J.N., Banks, R.E., Hutson, R., Harvey, D.J., Antrobus, R., Petrescu, S.M., Dwek, R.A., Rudd, P.M. (2007) Ovarian cancer is associated with changes in glycosylation in both acute-phase proteins and IgG. *Glycobiology* **17**(12):1344 – 1356.
doi: 10.1093/glycob/cwm100.

[5] Arnold, J.N., Saldova, R., Abd Hamid, U.M., Rudd, P.M. (2008) Evaluation of the serum *N*-linked glycome for the diagnosis of cancer and chronic inflammation. *Proteomics* **8**(16):3284 – 3293.
doi: 10.1002/pmic.200800163.

[6] Abd Hamid, U.M., Royle, L., Saldova, R., Radcliffe, C.M., Harvey, D.J., Storr, S.J., Pardo, M., Antrobus, R., Chapman, C.J., Zitzmann, N., Robertson, J.F., Dwek, R.A., Rudd, P.M. (2008) A strategy to reveal potential glycan markers from serum glycoproteins associated with breast cancer progression. *Glycobiology* **18**(12):1105 – 1118.
doi: 10.1093/glycob/cwn095.

[7] Arnold, J.N., Saldova, R., Galligan, M.C., Murphy, T.B., Mimura-Kimura, Y., Telford, J.E., Godwin, A.K. Rudd, P.M. (2011) Novel glycan biomarkers for the detection of lung cancer. *Journal of Proteome Research* **10**(4):1755 – 1764.
doi: 10.1021/pr101034t.

[8] Stöckmann, H., Adamczyk, B., Hayes, J., Rudd, P.M. (2013) Automated, high-throughput IgG-antibody glycoprofiling platform. *Analytical Chemistry* **85**(18):8841 – 8849.
doi: 10.1021/ac402068r.

[9] Campbell, M.P., Royle, L., Radcliffe, C.M., Dwek, R.A., Rudd, P.M. (2008) GlycoBase and autoGU: tools for HPLC-based glycan analysis. *Bioinformatics* **24**(9):1214 – 1216.
doi: 10.1093/bioinformatics/btn090.

Automated Detection and Identification of *N*- and *O*-Glycopeptides

Peter Hufnagel*, **Anja Resemann, Wolfgang Jabs, Kristina Marx, and Ulrike Schweiger-Hufnagel**

Bruker Daltonik GmbH, Fahrenheitstraße 4, 28359 Bremen, Germany

E-Mail: *peter.hufnagel@bdal.de

Received: 30th September 2013 / Published: 22nd December 2014

Abstract

Because it can provide detailed information about aglycons, glycosylation sites, and the composition and structure of glycans, mass spectrometry is highly suited for the analysis of glycopeptides and released *N*- and *O*-glycans. Here we present the bioinformatics platform ProteinScape, which can process entire LC-MS/MS runs, localise spectra that contain glycan-related information, and perform searches against glycan structure databases. An intuitive user interface facilitates interactive validation of results. If glycans are not released and glycopeptides are analysed, the heterogeneity of glycosylation at the various protein glycosylation sites can be assessed.

The integration of protein- and glycan-related functionality in a single software platform is particularly useful not only in glycoproteomics research, but also in biopharmaceutical development and QC. We provide several examples illustrating the efficiency of glycopeptide analysis using mass spectrometry. However, a comprehensive analysis requires information on the glycoprotein's mass profile. Therefore, the interpretation of mass spectra from intact glycoproteins is also discussed.

This article is part of the Proceedings of the Beilstein Glyco-Bioinformatics Symposium 2013.
www.proceedings.beilstein-symposia.org

Introduction

Glycosylation is one of the most common and important post-translational protein modifications. Glycoproteins have diverse functions and are involved in numerous biological processes. Their glycan moieties are either directly involved in regulatory processes or influence physicochemical properties of the glycoprotein. In eukaryotes, the vast majority of secreted and plasma membrane proteins are known to be glycosylated and glycosylation has been shown to be important for various protein functions [1]. The significance of glycoproteins is further demonstrated by the fact that more than 90% of proteins produced for therapeutic purposes – such as antibodies or hormones – are glycoproteins. In such biologics, glycosylation has been demonstrated to control parameters such as serum half-life or receptor interaction kinetics (e. g., [2]).

As a consequence, protein analysis projects – from large-scale proteomics down to focused protein characterisation studies – often require a detailed study of glycosylation. This is particularly true for the characterisation and quality control of recombinant biopharmaceutical proteins. However, due to a lack of dedicated and easy-to-use software, glycosylation is often neglected in the proteomics community. To address this issue, we have extended the scope of our bioinformatics software to the analysis of glycosylation.

ProteinScape is Bruker Daltonics' central bioinformatics platform for storage and processing of MS data. It supports various gel- and LC-based workflows. The underlying database organises all relevant data for all types of proteomics projects – including LC-data, gel data, mass spectra, process parameters and search results. ProteinScape acts as a central control unit and data evaluation tool for mass spectrometry-based identification, characterisation, and quantitation. The data hierarchy of the software is an accurate reflection of laboratory workflows – with projects, samples, separations, fractions; MS data and search results forming the main hierarchy levels. Information is accessible in tables and viewers and through extensive query functionalities. Dedicated viewers enable fast and interactive evaluation and validation of data and results (Figure 1). Real-time updating of the linked displays speeds up browsing and information management. As ProteinScape is able to handle data from different mass spectrometers, it is the ideal tool to study protein glycosylation.

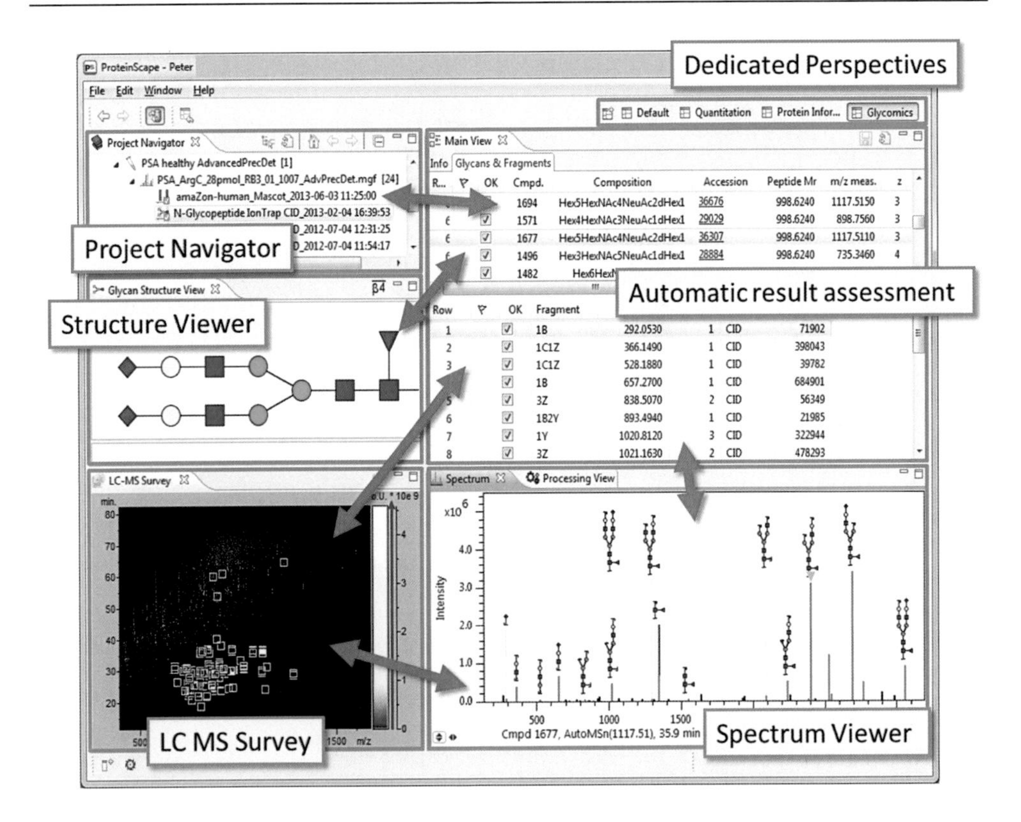

Figure 1. The ProteinScape user interface consists of closely linked navigators, tables, and data viewers. In addition to preset application-specific "Perspectives" (protein ID, protein quantitation, protein information, and glycomics), the arrangement of views can be freely adapted to the requirements of a specific workflow and saved as a user-defined "Perspective".

Three strategies for glycoprotein analysis

Three strategies can be applied for the comprehensive characterisation of a glycosylated protein. Figure 2 summarises these strategies.

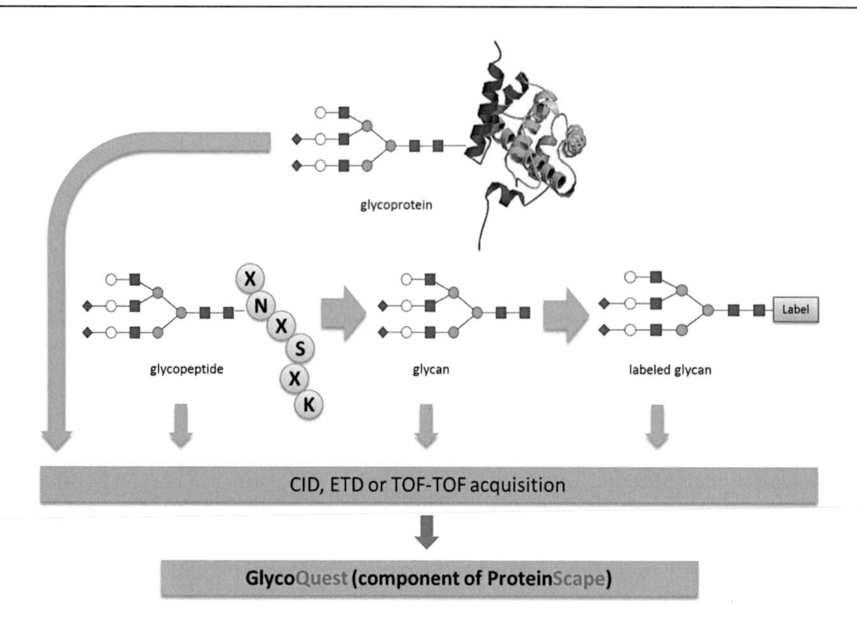

Figure 2. Glycoproteins can be analysed either as intact molecules, or after proteolytic digestion. Glycans can be analysed either as part of the respective glycopeptide, or as enzymatically- or chemically-released glycans that can be labelled using a fluorophore for optical detection in LC separations.

Strategy 1 is the separation of glycans from the protein or peptides. This common workflow generates a peptide fraction – that can be used for protein identification and the evaluation of other modifications – and a glycan fraction – that can be used for the identification of all glycans originally attached to the protein.

Compared to a proteolytic digest of a glycoprotein, the glycan fraction has a lower complexity and can be further separated and analysed by glycan-specific methods. The glycans can either be submitted to mass spectrometry without further modification, or a label (e. g., a fluorophore for visual detection) can be introduced to the reducing end of the glycans, which are then analysed by LC and/or mass spectrometry.

Strategy 2 is the proteolytic digestion of intact glycoproteins. If glycans are released, the information about the initial attachment of the respective glycans to the glycosylation sites of the individual proteins is lost. Proteolytic digestion of glycoproteins yields glycopeptides that can provide important information about the actual state of the various glycosylation sites.

Strategy 3 is the analysis of intact glycoproteins. Here, the heterogeneity of glycosylation is retained with the highest fidelity.

Strategy 1: Analysis of released glycans

ProteinScape's integrated processing pipeline contains several powerful algorithms: The GlycoQuest search engine is particularly useful for automated glycan analysis. Glycans released from the protein by a peptide-*N*-glycosidase (*N*-glycans) or by reductive beta-elimination (*O*-glycans) are automatically detected, identified, and reported.

As with protein database searches, glycan search parameters are defined and stored in a search method (Figure 3). Data are then searched against a user-defined glycan database or the meta-database GlycomeDB (http://www.glycome-db.org), which was developed at the German Cancer Research Center in Heidelberg and currently contains around 39,000 glycan structures. GlycomeDB is regularly updated with structures from various primary databases and automatically synchronised with GlycoQuest. Searches can be restricted to a primary database. A glycan structure editor enables manual editing of glycan structures that can be saved to a user-defined database for GlycoQuest searches. Figures 4 and 5 show two exemplary search results.

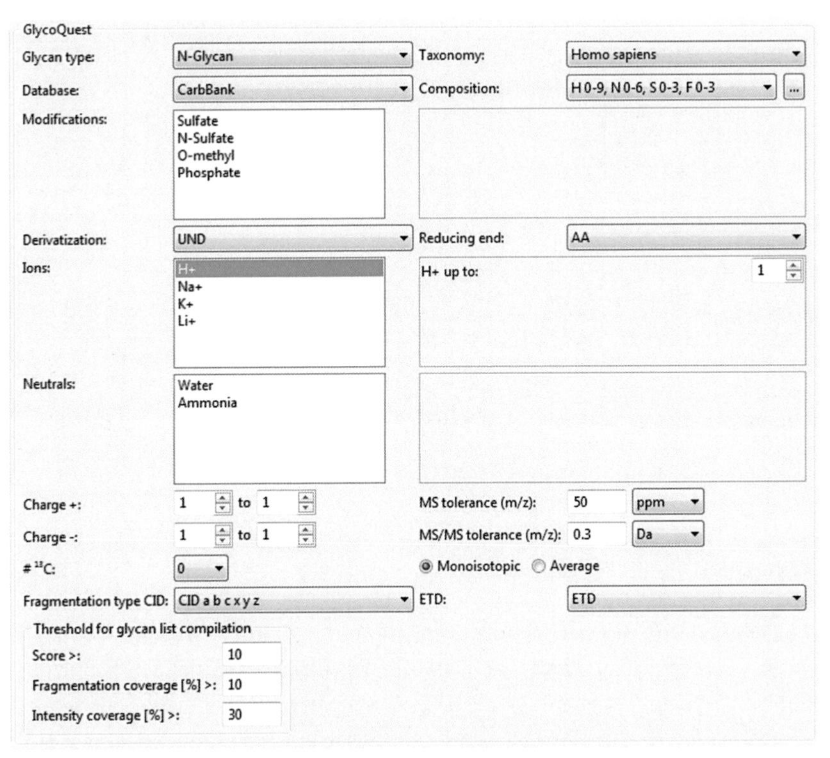

Figure 3. The GlycoQuest method editor. Parameters include one of GlycomeDB's primary databases (e. g. CarbBank), the glycan type (e. g. *N*-glycan), allowed compositions, and mass tolerances. Searches can be started for an individual spectrum, for an LC-MS/MS run, or for batches of spectra or LC-MS/MS runs.

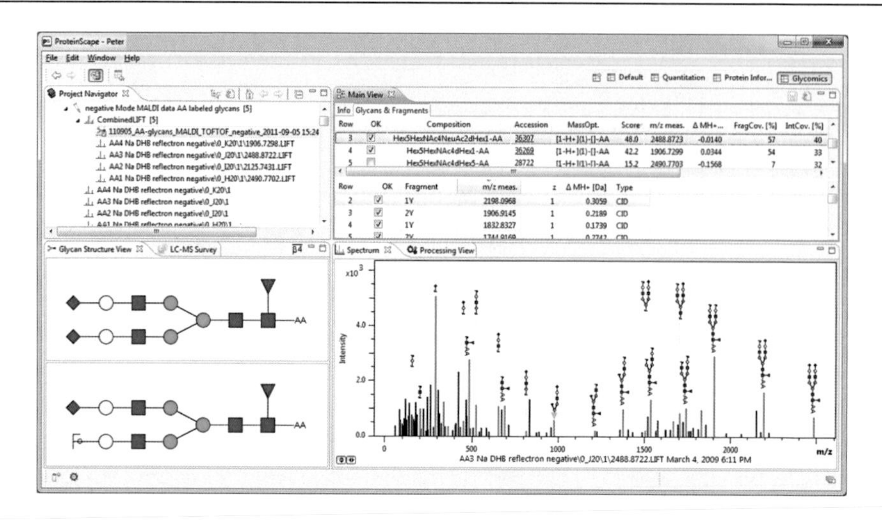

Figure 4. GlycoQuest result: a MALDI spectrum of a 2-aminobenzoic acid (AA)-labelled *N*-glycan, acquired on an ultrafleXtreme MALDI-TOF instrument.

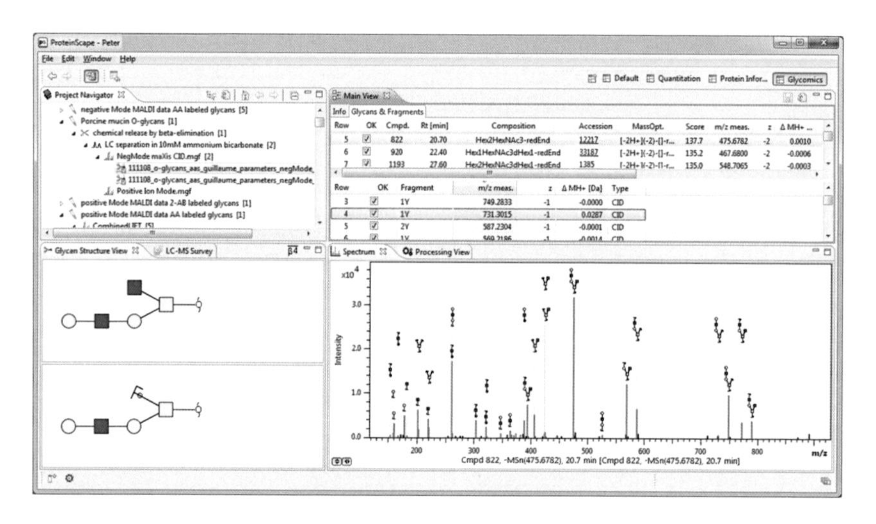

Figure 5. GlycoQuest result: an ESI spectrum of a reduced *O*-glycan, acquired on a maXis impact ESI-TOF instrument.

It should be noted here that a complete confirmation of a glycan's structure – which is highly dependent on the availability of all diagnostic fragment signals in the MS/MS spectrum – is a difficult task. GlycoQuest is an algorithm that compares spectra with theoretical fragment patterns. Some isomeric structures cannot be distinguished using this method. The situation becomes even more complicated if different glycans are not separated, and isobaric structures yield a mixture of fragments in the MS/MS spectrum. A satisfactory analysis of isomeric structures can be achieved by a variety of dedicated methods, such as permethylation of the sample, followed by the acquisition of MS^n spectra in an ion trap [3].

An alternative approach is the separation of reduced glycans by porous graphitised carbon chromatography. This technique is able to separate glycans that differ in structure but have the same composition and, therefore, the same mass. This allows differentiation of glycans with subtle linkage differences. For example, neuraminic acid [α2,3 or α2,6], galactose [β1,4 or α1,3] and the two G1 isomers that are frequently found on IgG *N*-glycans can be separated. In many cases the MS/MS spectra from such structures, although yielding identical theoretical fragments, have fundamental differences [4 – 7].

Strategy 2: Glycopeptide analysis

Glycopeptide analysis is a challenging task [8]. Due to high glycan heterogeneity and ion suppression effects, abundance of glycopeptide signals from proteolytic digests is usually low, and specific enrichment and separation techniques might be required. In addition, interpretation of MS/MS spectra is difficult as classical database search approaches cannot be used if the peptide's and the glycan's molecular weights are unknown. The correct determination of the peptide mass is a crucial feature for automated glycopeptide identification.

The glycopeptide classifier

The task of finding glycopeptide spectra in an LC-MS/MS dataset of a digested glycoprotein sample is demanding. ProteinScape searches for characteristic patterns in the MS/MS spectra and submits only relevant spectra to the database. Because both Mascot (Matrix Science) and GlycoQuest perform precursor-based searches, the exact masses of the glycan and peptide moieties of each glycopeptide must be known beforehand. The glycopeptide classifier algorithm of ProteinScape uses characteristic mass patterns (MALDI) or fragment ion series (ESI CID) to determine the glycan and peptide moiety masses. In this way, the glycopeptide classifier is mandatory for a successful identification of glycopeptides. Examples for *N*-glycopeptides are given in Figures 6 and 7. However, the glycopeptide classifier is not limited to *N*-glycopeptides. For example, core-fucosylated *N*-glycopeptides and also several kinds of *O*-glycopeptides in MALDI and ESI spectra can also be handled.

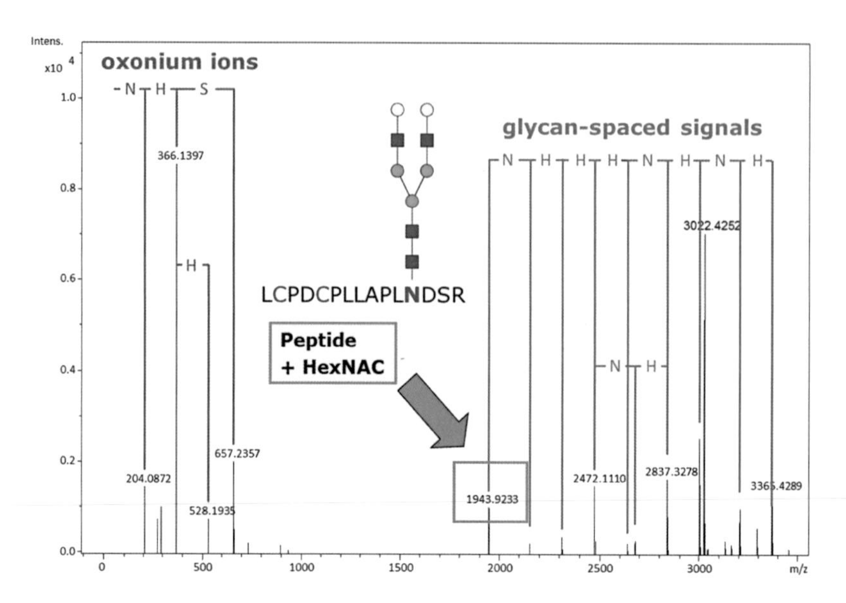

Figure 6. Deconvoluted ESI-CID spectra of *N*-glycopeptides usually contain diagnostic mass signals in the low molecular weight range, plus a characteristic tree of fragment ions that can be followed down to the mass of the peptide plus one *N*-acetylglucosamine.

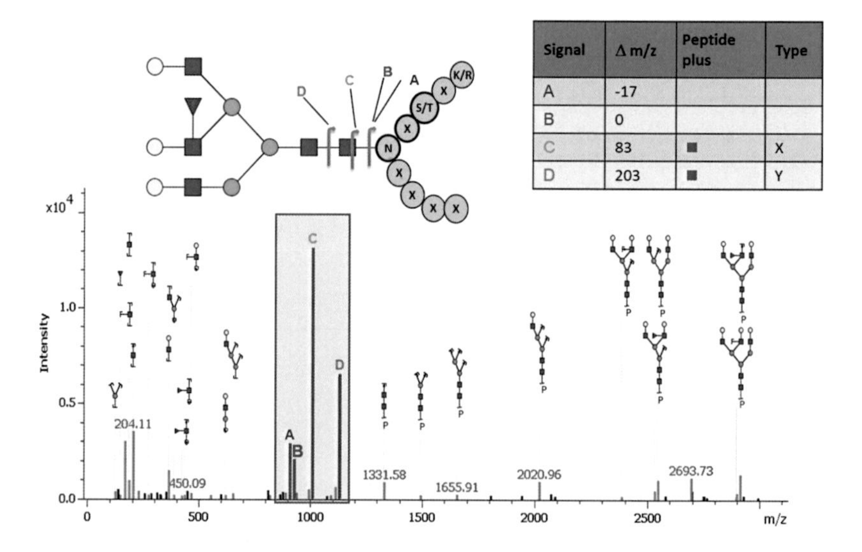

Signal	Δ m/z	Peptide plus	Type
A	-17		
B	0		
C	83	■	X
D	203	■	Y

Figure 7. MALDI-TOF/TOF spectra of *N*-glycosylated peptides contain a specific fragment pattern. A cleavage between the glycan and peptide part produces the MH+ of the peptide. The *N*-acetylglucosamine attached to the asparagine undergoes Y and ^{0,2}X cross-ring fragmentation providing strong fragment ions that are 83 and 203 Da heavier. A loss of ammonia from the asparagine completes the pattern ([8]).

Example: EPO

An example of the characterisation of a glycoprotein by the identification of glycopeptides is our study [9] on Erythropoietin (EPO). EPO is a glycoprotein with hormone activity that controls the production of red blood cells in bone marrow. Recombinant human EPO is produced on a large scale in cell culture as a therapeutic agent for treating anaemia related to different diseases. It is also abused as a blood doping agent in endurance sports. Human EPO is an approximately 34 kDa glycoprotein with one *O*- and three *N*-glycosylation sites (Figure 8). In pharmaceutical drug production, glycosylation heterogeneity of EPO is an important quality factor that influences functionality as well as bioavailability of the therapeutic protein.

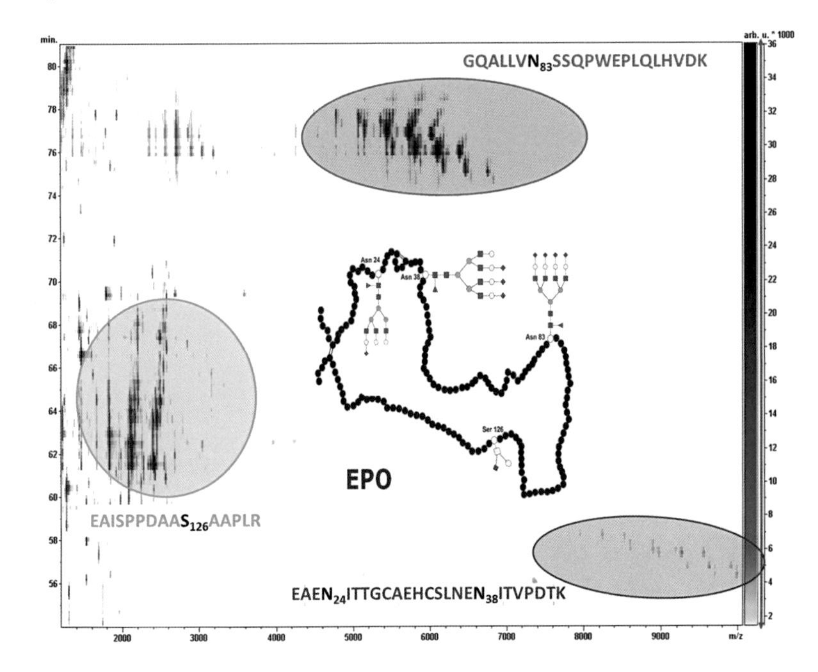

Figure 8. Survey view of an LC-MALDI-MS experiment on an EPO-BRP tryptic digest. Singly charged peptides (*m/z*) are displayed against retention time (min). Three glycopeptide fractions are visible: Two *N*-linked glycopeptide fractions (red and purple) and one *O*-linked glycopeptide fraction (green). The four glycosylation sites of EPO are one *O*-glycosylated serine (S 126), and three *N*-glycosylated asparagines (N24, N38 and N83). A particular difficulty arises from the fact that in a tryptic digest, N24 and N38 are found in the same peptide.

EPO BRP (LGC Standards) and recombinant human EPO expressed in HEK 293 cells (Sigma-Aldrich) were reduced, alkylated and subjected to trypsin digestion. The resulting peptides and glycopeptides were separated by nano-HPLC (nano-Advance, Bruker) and further analysed by MALDI-TOF/TOF-MS (ultrafleXtreme, Bruker). MS/MS spectra were imported into ProteinScape 3.1 and automatically screened for an *O*-glycopeptide-specific fragmentation pattern. The typical pattern for core 1 structures (– 18/+203/+162) was used to detect the relevant MS/MS spectra and to determine the peptide moiety mass of each glycopeptide. Subsequently, protein database searches were performed using Mascot (Matrix Science). GlycoQuest was used for glycan identification in GlycomeDB and in a custom-made EPO database that also contained acetylated structures not present in GlycomeDB (Figure 9). It could be shown that the glycosylation profile of EPO expressed in HEK 293 cells is much more heterogeneous than that of the BRP standard, and that the degree of acetylation is greatly reduced (Figure 10).

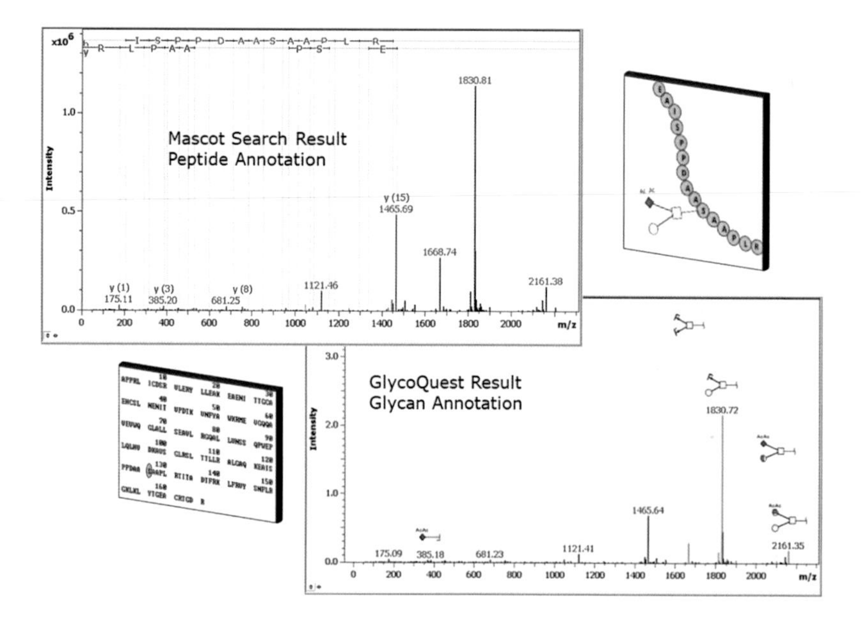

Figure 9. Mascot and GlycoQuest result for the spectrum of one *O*-glycopeptide. MALDI TOF/TOF spectra of *O*-linked glycopeptides contain both peptide and glycan fragments.

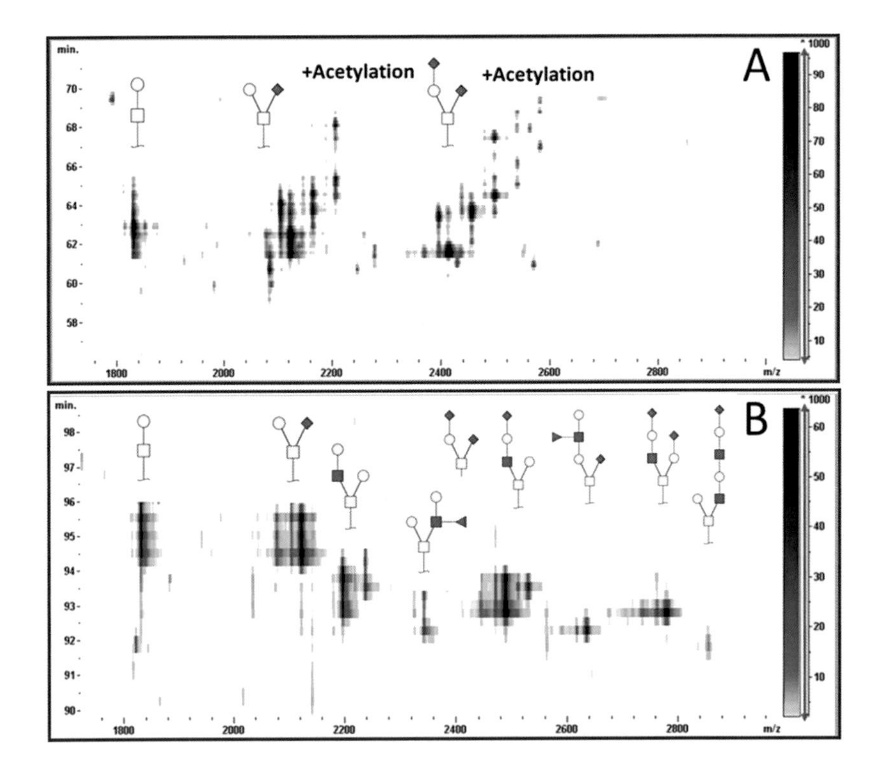

Figure 10. The Survey Viewer zoomed into the *O*-linked glycopeptide range of EPO BRP-Standard **(A)** and EPO HEK 293 cells **(B)** enables an at-a-glance comparison. Main differences were found in terms of glycan forms and acetylation grade.

Strategy 3: Glycoprotein analysis

Example 1: Cetuximab, a recombinant antibody

Because they play a central role in the immune response of vertebrates, antibodies represent one of the most important classes of glycoproteins. There is a growing interest in recombinant antibodies as biotherapeutic agents. Such so-called biologics are manufactured in biological systems, and usually employ recombinant technology. These systems can be very sensitive to minor changes in the manufacturing process, which may significantly alter the final biologic product. They can introduce modifications that may adversely affect the safety and efficacy of the drug. Therefore, strict quality control of each biopharmaceutical batch and comparison with reference standards is essential to ensure reproducibility between batches and to achieve regulatory approval.

Cetuximab is a chimeric mouse-human IgG1 that targets the epidermal growth factor receptor (EGFR). It is approved for use in the EU and US as a treatment for colorectal cancer and squamous cell carcinoma of the head and neck. The amino acid sequences for both the light and heavy chains of cetuximab (Figure 11) are reported in the IMGT database (http://www.imgt.org) and the drug bank (www.drugbank.ca). A high prevalence of hypersensitivity reactions to cetuximab were reported in some areas of the US. Different glycoforms were shown to be responsible for these hypersensitivity reactions and anaphylaxis. This example demonstrates the importance of a precise analysis of glycosylation in biopharmaceutical proteins.

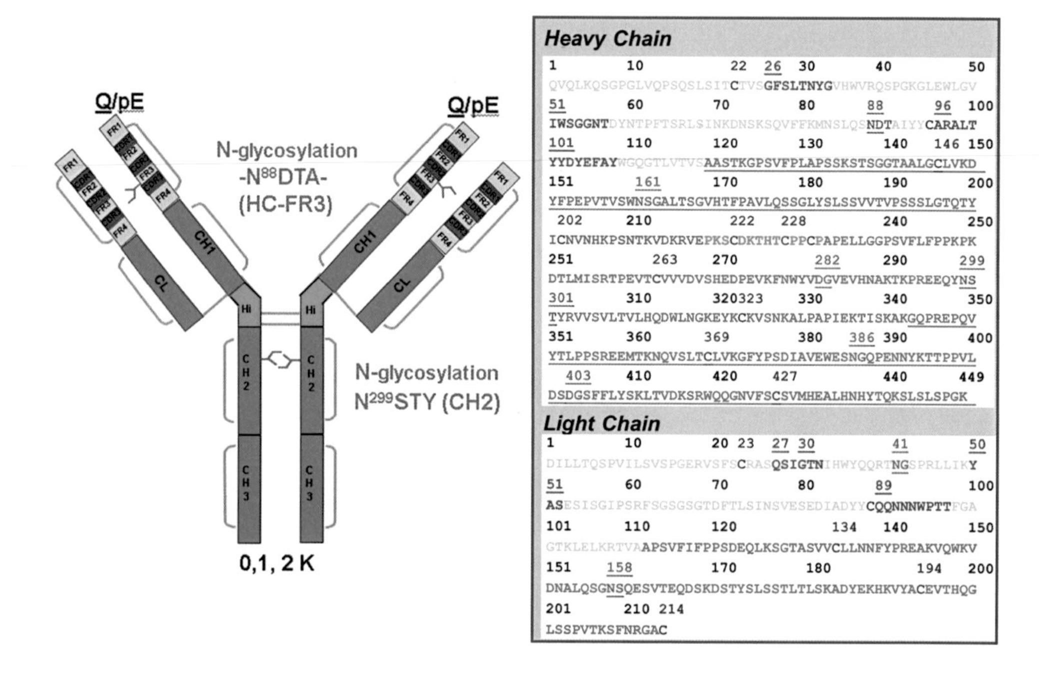

Figure 11. Schematic structure and sequence of cetuximab. Colours indicate protein QC "hot spots". Of particular importance are the two heavy-chain glycosylation sites, which can be separated from each other by proteolytic cleavage in the hinge region.

In the study of Ayoub *et al.* [10], detailed sequence information of the antibody's subunits was obtained using MALDI *N*- and *C*-terminal top-down sequencing (TDS) analysis. LC-MS/MS peptide mapping experiments on tryptic and GluC digests enabled post-translational modifications and sequence variants to be further localised.

For intact-mass analysis, the two heavy-chain *N*-glycosylation sites were separated by enzymatic cleavage (Figure 12). The LC-ESI mass spectra of the cetuximab subunits (middle-up approach) yielded glycosylation site-specific accurate masses of the various antibody glycoforms. Using GlycoQuest, glycopeptide and glycan identifications and profiles were automatically generated. Finally, the results from middle-up data were combined with the results from the bottom-up glycopeptide identification to generate a complete picture of the antibody's glycosylation (Figures 13 – 15).

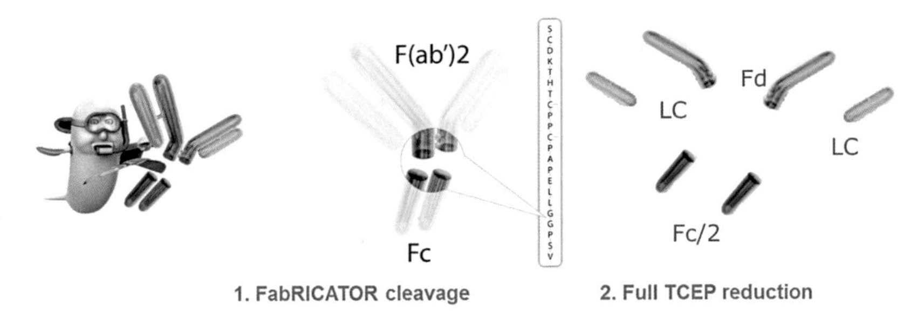

Figure 12. Middle-up approach: a dedicated proteolytic enzyme (FabRICATOR, Genovis) cleaves the heavy chain at the conserved Gly–Gly motif in the hinge region, thereby separating the two glycosylation sites [11], (Figure adapted from www.genovis.com).

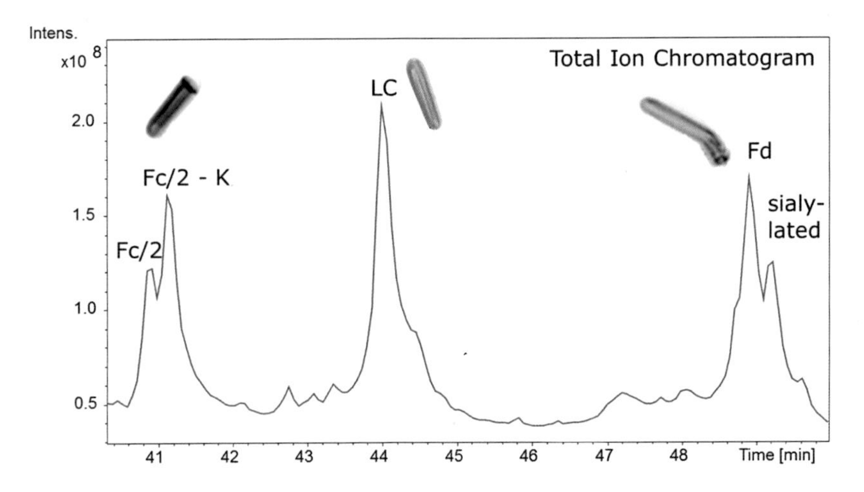

Figure 13. The light chain and both fragments of the heavy chain are separated by RP HPLC. The chromatogram shows a heterogeneous sialylation of the *N*-terminal Fd fragment (already indicating different glycosylation profiles for both sites) and lysine-clipping of the *C*-terminal Fc fragment, (Figure adapted from [10]).

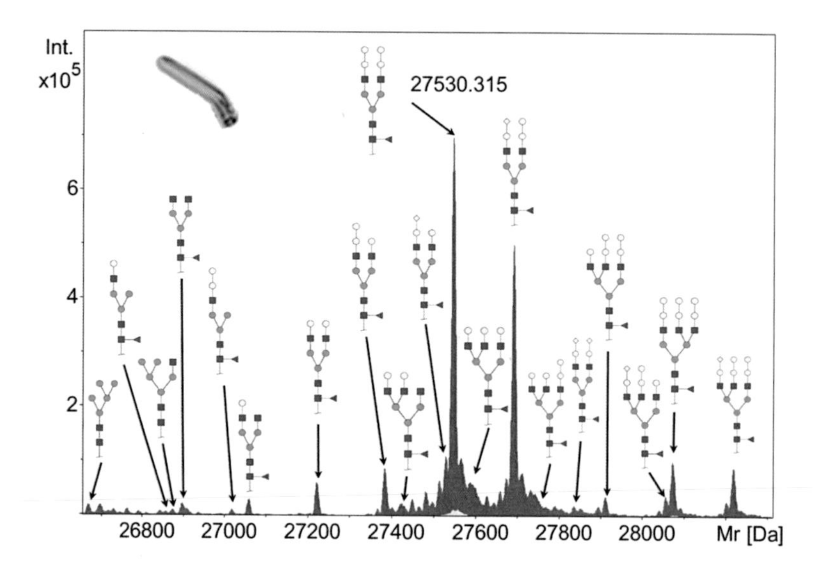

Figure 14. Glycosylation profile of the Fd fragment: Mass spectrum acquired during LC separation of the FabRICATOR fragments. Based on the "middle-up" data, GlycoQuest could assign the glycan compositions. Information about the glycan structures could be obtained in a separate analysis on the glycopeptide level ("bottom-up"), (Figure adapted from [10]).

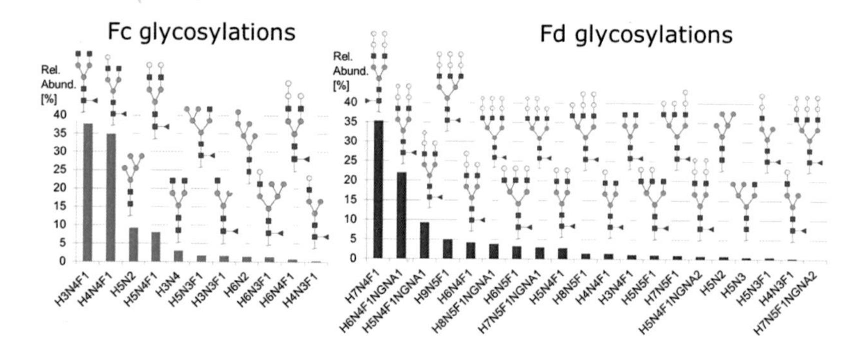

Figure 15. The different glycan profiles of Fc and Fd as determined by a two-way approach: Identification on the glycopeptide level ("bottom-up") plus quantitation on the FabRICATOR fragments ("middle-up"), (Figure adapted from [10]).

As pointed out in [11], the routine analysis of intact proteins or FabRICATOR-cleaved antibody subunits is a particularly useful technique for biopharmaceutical QC. Bruker's BioPharmaCompass software is designed for this task: After automated LC and data acquisition, detailed reports are automatically generated that show annotated total ion chromatograms and the annotated spectra of all key compounds. A simple traffic-light overview indicates which samples have passed the QC analysis, and which samples require re-investigation (Figure 16).

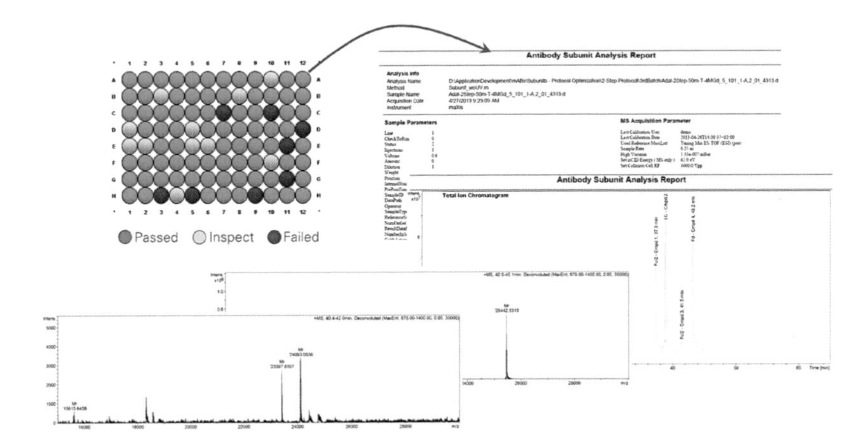

Figure 16. Automated report of BioPharmaCompass containing essential QC information such as TIC, deconvoluted protein mass and qualitative and quantitative comparison with a reference standard. Incorrect products and impurities are indicated using traffic-light colors for at-a-glance QC.

Example 2: PSA

The following results were generated during the 2013 gPRG study of the Glycoprotein Research Group (gPRG) of the Association of Biomolecular Resource Facilities (ABRF). The main objective of the study was the quantitation of the glycosylation heterogeneity present in two different preparations of human prostate-specific antigen (PSA), which is a biological biomarker for prostate cancer. A poster presented at the 2013 meeting of the ASMS focused on the identification of a new *N*-glycosylation site in the PSA study sample [12]. In brief, the glycosylation profiles of two PSA samples were analysed in a two-way approach: analysis of a tryptic digest by RP HPLC coupled to an amaZon Speed ion trap (Bruker) with CID and ETD capability ("bottom-up"), and analysis of the intact glycoprotein by RP HPLC coupled to a maXis 4G ESI-UHR-TOF (Bruker).

Because ion trap CID spectra of *N*-glycopeptides mainly contain peaks from glycan fragments (B- and Y-type ions) and peptide fragment peaks are rarely seen, little information about the peptide backbone is obtained. Therefore, both CID and ETD spectra were acquired. In contrast to CID, ETD cleaves the *N*-Cα bond of the peptide backbone, resulting in c- and z-ion series. Post-translational modifications – for example glycans – remain attached to the respective amino acid residues. An example is shown in Figure 17. More information on ETD of glycopeptides can be found in 13]. As a result, 50 glycan compositions could be assigned to the glycan profile with quantitative information. From 44 glycans, the structure could be deduced from "bottom-up" CID spectra (Figure 18).

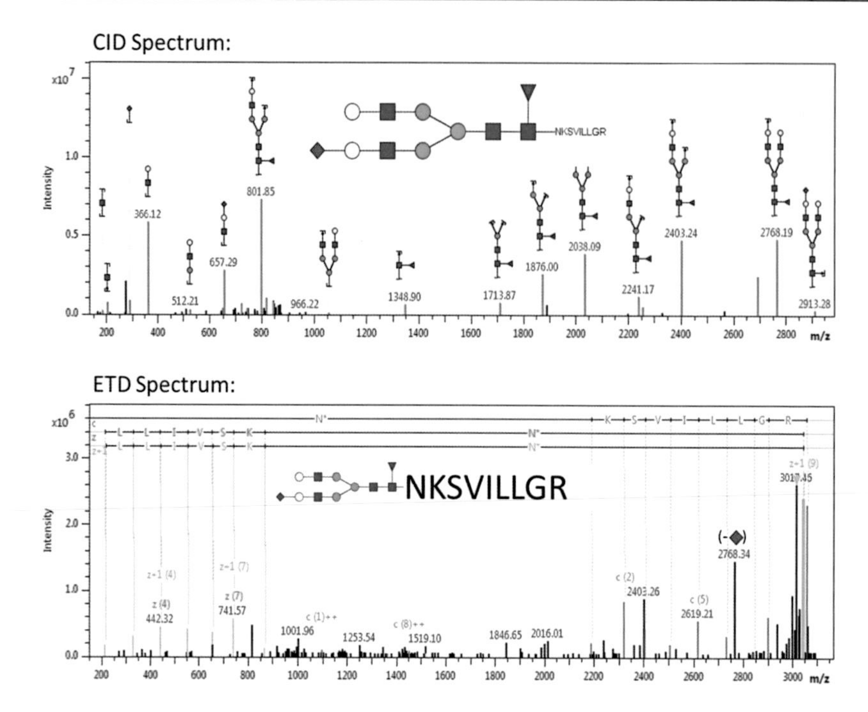

Figure 17. CID and ETD spectrum of a single glycopeptide derived from PSA. The composition of the glycan moiety can be identified by submission of the CID spectrum to GlycoQuest. The peptide moiety can be identified by submission of the ETD spectrum to Mascot once the glycan composition is determined.

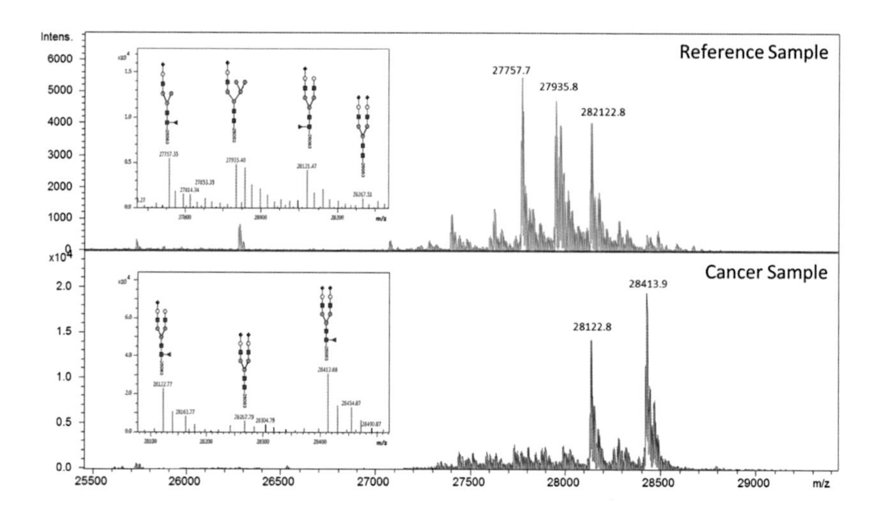

Figure 18. Intact mass spectra of the two PSA samples, annotated with the glycan structures identified from the "bottom-up" approach.

MS of intact glycoproteins is highly suited to the quantitative assessment of glycoprofiles. Because the glycans carry a huge protein residue, suppression effects familiar from complex glycan or glycopeptide spectra, are eliminated [14]. In addition, this approach avoids artefacts that can arise from differences in charge-state distributions or incomplete or unspecific proteolytic digestion. However, only glycoproteins with a single glycosylation site can be quantified without further fragmentation of the protein.

Conclusion

The characterisation of glycoproteins is a complex task. In addition to the complete assignment of the amino-acid sequence and the analysis of other post-translational modifications (e.g., phosphorylation) and laboratory artefacts (e.g., oxidation), the investigation of glycosylation and its protein-specific heterogeneity adds further analytical challenges. Because many proteins contain several potential glycosylation sites, and due to a high degree of glycan heterogeneity at a given amino acid, glycosylation patterns are often very complex. Three strategies for the analysis of glycoproteins have been described here:

1. Analysis of released glycans: Reduced complexity, lower mass range, but only very limited information about the glycosylation site(s). Porous graphitised carbon chromatography can separate isobaric glycans.
2. Analysis of glycopeptides in a proteolytic digest: The link to the glycosylation site is retained. However, glycoprotein digests are complex mixtures, and in many cases a specific glycopeptide enrichment step is necessary.
3. Analysis of the intact glycoprotein: Yields the exact quantitative glycosylation profile. Not suitable for proteins with several glycosylation sites.

In many cases a complete overview of the protein's actual state of glycosylation can be produced only by the combination of two or even all three strategies. We have introduced dedicated bioinformatics software that supports each single strategy and significantly facilitates a combined approach.

References

[1] Kolarich, D., Lepenies, B., Seeberger, P.H. (2012) Glycomics, glycoproteomics and the immune system. *Current Opinion in Chemical Biology* **16**:214 – 220. doi: 10.1016/j.cbpa.2011.12.006

[2] Sinclair, A.M., Elliott, S. (2005) Glycoengineering: the effect of glycosylation on the properties of therapeutic proteins. *Journal of Pharmaceutical Sciences* **94**: 1626 – 1635. doi: 10.1002/jps.20319

[3] Hailong Zhang, David J. Ashline and Vernon N. Reinhold (2014) Tools to MSn Sequence and Document the Structures of Glycan Epitopes. In: Proceedings of the 3rd Beilstein Glyco-Bioinformatics Symposium 2013 (Eds. Hicks, M.G. and Kettner, C.). Logos-Verlag Berlin, pp 117 – 131.

[4] Everest-Dass, A.V., Abrahams, J.L., Kolarich, D., Packer, N.H., Campbell, M.P. (2013) Structural feature ions for distinguishing *N*- and *O*-linked glycan isomers by LC-ESI-IT MS/MS. *Journal of The American Society for Mass Spectrometry* **24**:895 – 906. doi: 10.1007/s13361-013-0610-4

[5] Jensen, P.H., Karlsson, N.G., Kolarich, D., Packer, N.H. (2012) Structural analysis of *N*- and *O*-glycans released from glycoproteins. *Natural Protocols* **7**:1299 – 1310. doi: 10.1038/nprot.2012.063

[6] Stadlmann, J., Pabst, M., Kolarich, D., Kunert, R., Altmann, F. (2008) Analysis of immunoglobulin glycosylation by LC-ESI-MS of glycopeptides and oligosaccharides. *Proteomics* **8**:2858 – 2871. doi: 10.1002/pmic.200700968

[7] Kolarich, D. (2014) In: Proceedings of the 3rd Beilstein Glyco-Bioinformatics Symposium 2013 (Eds. Hicks, M.G. and Kettner, C.). Logos-Verlag Berlin.

[8] Wuhrer, M., Catalina, M.I., Deelder, A.M., Hokke, C.H. (2007) Glycoproteomics based on tandem mass spectrometry of glycopeptides. *Journal of Chromatography B* **849**:115 – 128. doi: 10.1016/j.jchromb.2006.09.041

[9] Resemann, A., Tao, N., Schweiger-Hufnagel, U., Marx, K., Kaspar, S. (2013) Comprehensive Study of *O*-Linked Glycans of Erythropoietin. ASMS 2013. ThP19 356, available at http://www.bruker.com/products/mass-spectrometry-and-separations/literature/literature-room-mass-spec.html (section search: "Erythropoietin").

[10] Ayoub, D., Jabs, W., Resemann, A., Evers, W., Evans, C., Main, L., Baessmann, C., Wagner-Rousset, E., Suckau, D., Beck, A. (2013) Correct primary structure assessment and extensive glyco-profiling of cetuximab by a combination of intact, middle-up, middle-down and bottom-up ESI and MALDI mass spectrometry techniques. *mAbs* **5**:699 – 710.
doi: 10.4161/mabs.25423

[11] Olsson, F., Andersson, L., Willetts, M., Jabs, W., Resemann, A., Evers, W., Baessmann, C., Suckau, D. (2013) Optimizing the Enzymatic Subunit Generation with IdeS for High Throughput Structure Verification of Therapeutic Antibodies by Middle-Down Mass Spectrometry. ASMS, WP24 – 434, available at http://www.bruker.com/products/mass-spectrometry-and-separations/literature/literature-room-mass-spec.html (section search: "wp24 – 434").

[12] Schweiger-Hufnagel, U., Marx, K., Jabs, W., Resemann, A. (2013) Qualitative and quantitative investigation of glycans attached to Prostate-specific antigen (PSA) glycoprotein of healthy and cancer samples. ASMS 2013, ThP 19 – 339, available at http://www.bruker.com/products/mass-spectrometry-and-separations/literature/literature-room-mass-spec.html (section search: "glycoprotein") and http://www.abrf.org/Other/ABRFMeetings/ABRF2013/RG presentations/RG14_gPRG_Leymarie.pdf.

[13] Marx, K., Kiehne, A., Meyer, M. (2013) amaZon speed ETD: Exploring glycopeptides in protein mixtures using Fragment Triggered ETD and CaptiveSpray nanoBooster, Bruker Application Note LCMS-93, available at http://www.bruker.com/products/mass-spectrometry-and-separations/literature/literature-room-mass-spec.html (section search: "glycopeptide").

[14] Stavenhagen, K., Hinneburg, H., Thaysen-Andersen, M., Hartmann, L., Varón Silva, D., Fuchser, J., Kaspar, S., Rapp, E., Seeberger, P.H., Kolarich, D. (2013) Quantitative mapping of glycoprotein micro- and macro-heterogeneity: An evaluation of mass spectrometry signal strengths using synthetic peptides and glycopeptides. *Journal of Mass Spectrometry* **48**:627 – 639.
doi: 10.1002/jms.3210

Discovering the Subtleties of Sugars
June 10th – 14th, 2013, Potsdam, Germany

New Structure–function Relationships of Carbohydrates

Thisbe K. Lindhorst

Otto Diels Institute of Organic Chemistry, Christiana Albertina University of Kiel
Otto-Hahn-Platz 3 – 4, D-24118 Kiel, Germany

E-Mail: tklind@oc.uni-kiel.de

Received: 5th May 2014 / Published: 22nd December 2014

Abstract

The potential of glycoarrays for the investigation of carbohydrate interactions has not been fully exploited to date. In addition to the saccharide specificity of lectins, carbohydrate recognition and carbohydrate binding most likely also comprises aspects of pattern formation, density regulation, as well as the mode of sugar presentation on a surface. When glycoarrays – which allow for systematic alteration of such parameters – become available, new structure–function relationships are likely to be discovered in the carbohydrate regime. In this account some of our work on fabrication of special glycoarrays is summarised including the 'dual click approach' to glyco-SAMs, and fabrication of photosensitive glycoarrays which allow 'switching' of carbohydrate orientation between two distinct states.

Introduction

Organismic life has been partitioned into life of cells and cell–cell interactions, and hence, cell biology has emerged as an own scientific discipline during the late 20th century. However, till this date, it has been greatly overlooked, that the glycosylated cell surface ('glycocalyx') is a key feature of all cells, and of eukaryotic cells in particular. In addition, it has been frequently underestimated, how important the carbohydrates as glycocalyx main components are for cell biology and as regulators of cell–cell interactions. Likewise, the meaning of molecular diversity of the carbohydrates remains insufficiently understood in comparison to the two other major biopolymer classes, the nucleic acids and the proteins.

This article is part of the Proceedings of the Beilstein Glyco-Bioinformatics Symposium 2013.
www.proceedings.beilstein-symposia.org

The growing understanding of DNA and protein functions has led to meaningful research fields, namely the genomics and the proteomics. These are given priority in biochemistry, whereas the 'glycomics' [1] rather receive second-rate treatment. A possible explanation for this situation lies in the overwhelming complexity of carbohydrate chemistry as well as of carbohydrate biochemistry, which is indeed difficult to manage.

Thus, a cell's glycocalyx appears like a molecular puzzle to researchers, which needs to be unravelled to understand the meaning of carbohydrates for life. Apparently, the lectins, a huge class of carbohydrate-recognizing proteins, contribute critically to glycobiology through the formation of specific carbohydrate–protein interactions. Consequently, a great deal of attention has been paid to experimental formats that allow systematic as well as versatile testing of carbohydrate–protein interactions. Glycoarrays have emerged as number one tool in this regard [2 – 5]. Glycoarrays are artificial glycosylated surfaces in which more or less complex oligosaccharides are immobilised on a suitable material of different dimension. Experiments with glycoarrays typically involve fabrication of glycoarrays on the one hand and their interrogation with lectins on the other. Thereby, much valuable information about the specificity of lectins has been generated and related to the structures of complex carbohydrates [6,7].

Pride and Prejudice of Glycoarrays

While the approach of probing sophisticated glycoarrays with proteins has clearly increased our understanding of carbohydrate–lectin interactions, a number of questions in addition to the aspects of sugar specificity of recognition, remain open as highlighted in Figure 1. Two fundamental problems of glycoarray fabrication are illustrated, namely (i) the regulation of density of carbohydrates on a surface, and (ii) the challenges connected to fabrication of mixed (thus more divers) glycoarrays, with both topics related to the possibility of patterning or raft formation, respectively. The requirement of uniform distribution of different carbohydrates on a surface to form a regularly mixed glycoarray is difficult to fulfil as well as the success of any such attempt is hard to analyse. Segregation effects occurring on a surface regardless of whether one or more types of carbohydrates were immobilised, normally remain undetected and consequently, even misinterpretation of testing results might happen.

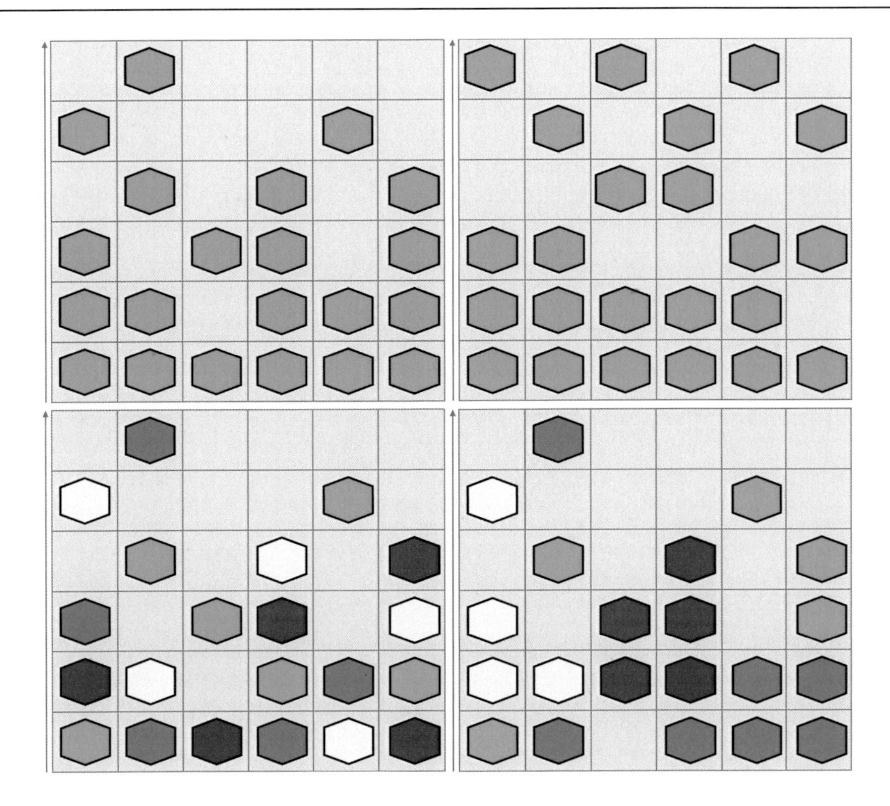

Figure 1. Density of carbohydrates on a surface (top) and regular mixing of different sugars in glycoarray formation (bottom) are difficult to control as well as to analyse. Segregation effects (on the right) might occur in all cases regardless of whether a pure or a mixed glycoarray is prepared. Hexagons represent carbohydrates (not further specified). Arrows on the left next to the depicted surface squares indicate increasing dilution of sugar solutions that are applied for glycoarray fabrication.

A lot of our own research has been dedicated to the investigation of bacterial adhesion, in particular adhesion of uropathogenic *Escherichia coli*, in short UPEC [8 – 11]. This work bears the potential for advancement of our understanding of the mechanisms of cell adhesion and, secondly, the possibility of progress towards an 'anti-adhesion therapy' against microbial infection. Adhesion of bacteria cells is mediated, inter alia, by extracellular adhesive organelles, called fimbriae that carry lectin domains to attach the bacterial cell to the glycosylated cell surface of a target cell through multiple carbohydrate-(bacterial) lectin interactions. For UPEC infections, the so-called type 1 fimbriae are of particular relevance, mediating α-D-mannoside-specific adhesion [12]. Consequently, type 1 fimbriae-mediated bacterial adhesion can be inhibited by α-D-mannosides or suitable antagonists. Based on the structure of the type 1 fimbrial lectin, the protein FimH, a large number of carbohydrate inhibitors of mannose-specific bacterial adhesion has been invented and investigated in different assays [13, 14]. Importantly, in addition to using inhibitors of bacterial adhesion in solution, we have started testing various designer surfaces, in other words glycoarrays, to test bacterial adhesion.

Our bacterial adhesion experiments often involve the fabrication of glycoarrays using a microtiter plate format. For this, differently concentrated carbohydrate solutions are employed for different rows of wells. In this procedure, it is assumed that the carbohydrates density of surface coverage can be systematically varied across the microplate wells (Figure 1). Typically, this procedure of concentration variation reveals an optimal carbohydrate concentration for a particular cell adhesion experiment, resulting in a maximally adhesive surface, while further dilution of the carbohydrate solutions that are employed for glycoarray fabrication leads to less dense glycoarrays and consequently less adhesion of (bacterial) cells.

Density Control of Glycoarrays

Recently, we have obtained indications, that the clustering of carbohydrates, such as α-D-mannosyl moieties, in *one* molecule resulting in so-called cluster glycosides might be used as means to control the carbohydrate density of glycoarrays [15]. In this experiment a series of simple (monovalent) mannosides (Figure 2, **1**) and their di- and trivalent glycocluster analogues (**2** and **3**) were employed.

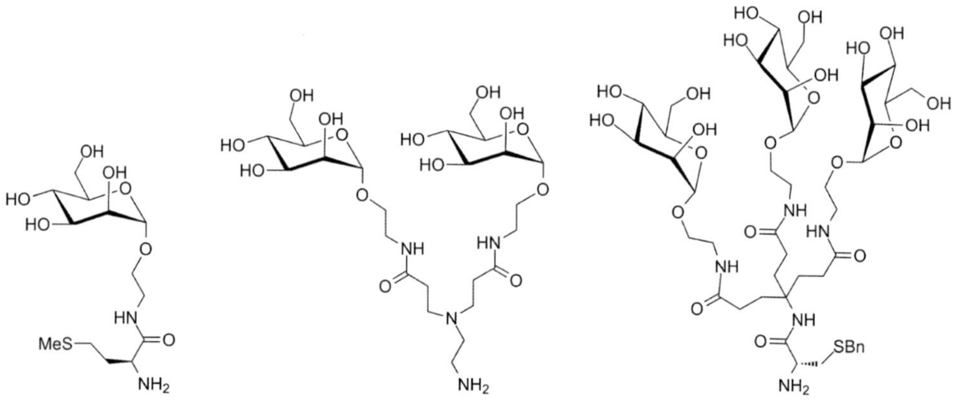

Figure 2. Mannoside **1**, and its di- and trivalent analogues **2** and **3** were prepared to test density carbohydrate control of glycoarrays [15].

Testing of bacterial adhesion using pre-functionalised microtiter plates and solutions of **1**, **2**, and **3**, respectively, to fabricate the respective glycoarrays led to a picture as reflected in Figure 3. It was shown that at higher concentrations (~15 mmol), all three types of glyco-arrays are approximately equally adhesive within the experimental error. On the contrary, at lower concentrations (~2 mmol), the glycoarray prepared from a solution of the trivalent cluster mannoside **3** became more adhesive in comparison to the respective glycoarrays made from **1** or **2** in the same concentration range.

The interpretation of the results was that each of the three types of employed glycoarrays, prepared from **1**, **2**, or **3**, respectively, shows different concentration dependencies in bacterial adhesion assays. When higher sample concentrations were used for the immobilisation, density of mannose coverage should be comparable in all three cases. On the other hand, further dilution of the carbohydrate solutions employed for surface functionalisation would affect carbohydrate density more critically in case of the glycoarrays resulting from **1**, whereas glycoarrays prepared from the trivalent cluster glycoside **3** can still provide relatively high local α-mannoside density (Figure 3, right).

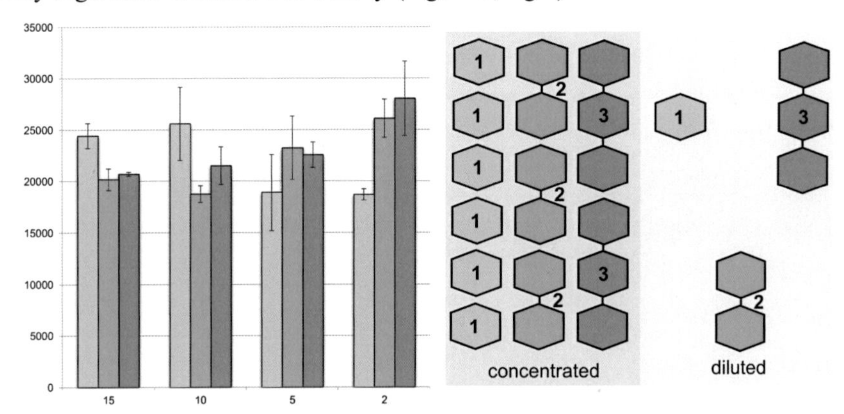

Figure 3. Three types of glycoarrays were prepared using pre-functionalised microtiter wells and solutions of 1 (Figure 2, left bars), 2 (middle bars), or 3 (right bars), respectively, at concentrations of 15, 10, 5, and 2 mmol (x axis). Concentrations are valency-corrected (that means that the specified concentrations refer to α-D-mannosyl moieties rather than to the molecular concentration of the solution that was used for glycoarray fabrication). Adhesion of fluorescent *E. coli* cells was determined in each case by fluorescence read-out (standard deviations are indicated for each bar) (y axis). For interpretation of these findings see the cartoon on the right: When higher sample concentrations were used for the immobilisation, density of mannose coverage should be comparable in all three cases. On the other hand, dilution of the carbohydrate solutions employed for surface functionalisation should be more critical regarding the resulting carbohydrate density in case of the glycoarrays resulting from 1 than those fabricated from 3.

In the near future it has to be investigated if this or other alternative concepts [16 – 17] for the control of carbohydrate density on surfaces can be advanced into a reliable methodology. In parallel, we have worked on methods for the consecutive build-up of glycoarrays to increase our options in making and manipulating glycosylated surfaces. Thus, we have recently expanded the repertoire of glycoarray fabrication by a ‘dual click approach’ in which glycoarrays were constructed as ‘glyco-SAMs’ using three simple consecutive steps.

Dual Click Approach to Glyco-SAMs

The term SAM stands for self-assembled monolayer, founded by Whitesides and co-workers in the 1980s [18]. Typical SAMs are fabricated by the reaction of an alkanethiol on a gold surface through formation of Au–S bonds, leading to rather well-organised, regular monolayers that are amenable to a number of (bio)physical experiments comprising an impressive explanatory power [19, 20]. Moreover, glyco-SAMs have been shown to be of great value in the glycosciences including the testing of bacterial adhesion. In the dual click approach to glyco-SAM fabrication we have employed long chain mercapto-alkynes for the assembly of a principal, alkyne-terminated SAM, followed by a first 'click reaction' involving Cu(I)-catalysed coupling to α-azido-ω-amino-difunctionalised oligoethylene glycol (OEG) on SAM [21]. This step provides 'biorepulsive' properties of the SAM, this is, non-specific adhesion of proteins to this surface is prohibited through the OEG unit. Next, in the second 'click reaction' the terminal amino group can be further refined on SAM by employing NCS-functionalised bio-molecules such as glycosyl isothiocyanates, or other isothiocyanato-modified glycosides or glycoclusters, respectively (Figure 4). We could show that the prepared glyco-SAMs provide suitable platforms to test bacterial adhesion.

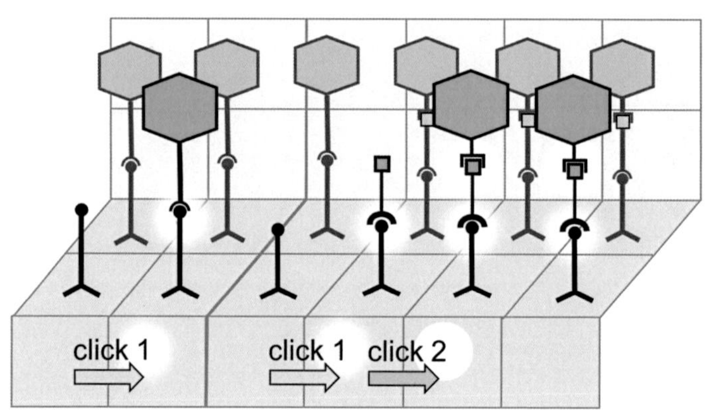

Figure 4. A 'dual click' approach allows consecutive assembly of glyco-SAMs 'on SAM' [21].

Photosensitive Glycoarrays

An addition to aspects of glycoarray fabrication with regard to the problems of density control and consecutive build-up, we have become interested in glycosylated surfaces, which allow for controlled manipulation of carbohydrate exposition; ideally with spatiotemporal resolution. This is of importance in the context of conformational control as a regulatory mechanism in glycobiology. Conformational control of biological function is well-known and highly appreciated in other areas of biochemistry, for example in structural biology. In glycobiology it has remained a neglected area to date.

Searching for glycoarrays that can be switched between to different steric states without changing other parameters of the glycosylated surface (except carbohydrate orientation), we became interested in azobenzene glycosides [22]. The photochemical *E*/*Z* isomerisation of the azobenzene N = N double bond can be easily achieved by appropriate irradiation and has been established as a biocompatible method for photocontrol in biology [23, 24]. Thus, the azobenzene N = N double bond might be utilised as a hinge region in an azobenzene glycoconjugate, permitting controlled steric manipulation of a glycoarray as outlined in Figure 5.

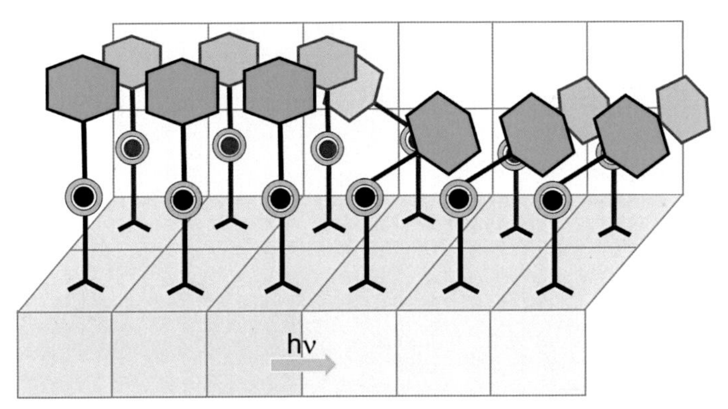

Figure 5. Azobenzene glycoarrays should in principle be isomerisable using light of appropriate wavelength. The azobenzene N = N double bond serves as a hinge region. The effected steric changes are reversible as E→Z isomerisation requires a different wavelength than Z→E back isomerisation. Such 'photo switching' on surfaces can be applied to probe conformational control of carbohydrate recognition.

We have shown that azobenzene glycosides as those depicted in Figure 6 can be readily synthesised [25 – 27], possess favourable photochromic properties and are biocompatible and non-toxic [28]. Moreover, azobenzene derivatives including azobenzene glycosides can be elaborated into molecular tools for bioorthogonal ligation chemistry [29]. Thus, azobenzene glycosides are emerging into a class of photosensitive glycoconjugates that are suited to test conformational control of carbohydrate recognition. However, the question remains if photocontrol is effective on glycosylated surfaces. Thus, we have employed IRRAS (infrared reflection absorption spectroscopy) for the characterisation of glyco-SAMs. IRRAS allows to measure vibrational changes of films or molecular monolayers on surfaces and is therefore suited to determine if photoisomerisation of chemisorbed azobenzene glycosides (Figure 7) [30].

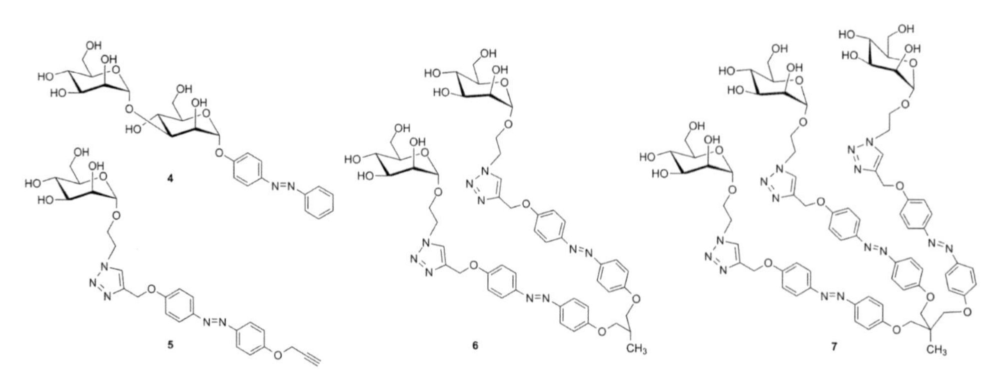

Figure 6. Examples of synthetic mono-, di- and trivalent azobenzene glycoconjugates.

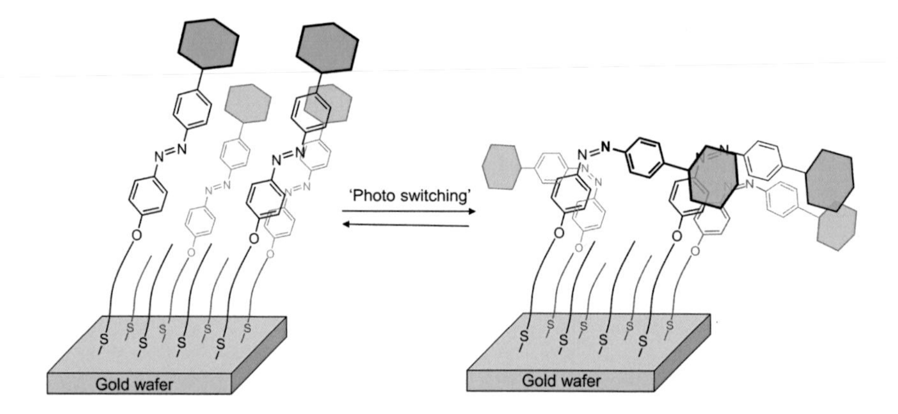

Figure 7. Glyco-SAMs were fabricated using mercapto-functionalised azobenzene glycoside derivatives. Photoswitching was proven by IRRAS [30].

To this end, it was shown that photosensitive glycoarrays can be prepared and their carbohydrate orientation can indeed be manipulated using light of appropriate wavelength (~365 nm). The next step in this project will be testing of cellular adhesion in relation to carbohydrate orientation on a surface.

Conclusion

The glycosylated surface of cells is probably one of the most exciting surfaces known. Compared to the surface of our planet it is small, but its complexity and its potential for molecular diversity might be comparable to mother earth. Glycoarrays are useful tools to investigate features and functions of glycosylated surfaces but many adventures lie ahead of researchers who try to expand the potential of glycoarrays. So far it remains difficult to control patterning of glycoarrays and density of carbohydrate coverage as well as to enable controlled manipulation of glycoarray surfaces. This account is meant to inspire further research and discoveries in this area.

Acknowledgements

Financial support by the collaborative network SFB 677 (DFG) is gratefully acknowledged.

References

[1] Rakus, J.F., Mahal, L.K. (2011) New Technologies for Glycomic Analysis: Toward a Systematic Understanding of the Glycome. *Annual Review of Analytical Chemistry* **4**:367 – 392.
doi: 10.1146/annurev-anchem-061010-113951.

[2] Turnbull, J.E., Field, R.A. (2007) Emerging glycomics technologies. *Nature Chemical Biology* **3**:74 – 77.
doi: 10.1038/nchembio0207-74.

[3] Feizi, T., Fazio, F., Chai, W., Wong, C.-H. (2003) Carbohydrate microarrays – a new set of technologies at the frontiers of glycomics. *Current Opinion in Structural Biology* **13**:637 – 645.
doi: 10.1016/j.sbi.2003.09.002.

[4] Laurent, N., Haddoub, R., Voglmeir, J., Wong, S.C.C., Gaskell, S.J., Flitsch, S.L. (2008) SPOT synthesis of peptide arrays on self-assembled monolayers and their evaluation as enzyme substrates. *ChemBioChem* **9**:2592 – 2596.
doi: 10.1002/cbic.200800481.

[5] Horlacher, T., Seeberger, P.H. (2008) Carbohydrate arrays as tools for research and diagnostics. *Chemical Society Reviews* **37**:1414 – 1422.
doi: 10.1039/B708016F.

[6] Rillahan, C.D., Paulson, J.C. (2011) Glycan Microarrays for Decoding the Glycome. *Annual Review of Biochemistry* **80**:797 – 823.
doi: 10.1146/annurev-biochem-061809-152236.

[7] von der Lieth, C.-W., Freire, A.A., Blank, D., Campbell, M.P., Ceroni, A., Damerell, D. R., Dell, A., Dwek, R.A., Ernst, B., Fogh, R., Frank, M., Geyer, H., Geyer, R., Harrison, M.J., Henrick, K., Herget, S., Hull, W.E., Ionides, J., Joshi, H.J., Kamerling, J.P., Leeflang, B.R., Lütteke, T., Lundborg, M., Maass, K., Merry, A., Ranzinger, R., Rosen, J., Royle, L., Rudd, P.M., Schloissnig, S., Stenutz, R., Vranken, W. F., Widmalm, G., Haslam, S.M. (2011) EUROCarbDB: An open-access platform for glycoinformatics. *Glycobiology* **21**:493 – 502.
doi: 10.1093/glycob/cwq188.

[8] Grabosch, C., Kolbe, K., Lindhorst, T.K. (2012) Glycoarrays by a new tandem noncovalent-covalent modification of polystyrene microtiter plates and their interrogation with live cells. *ChemBioChem* **13**:1874 – 1879.
doi: 10.1002/cbic.201200365.

[9] Wehner, J.W., Weissenborn, M.J., Hartmann, M., Gray, C.J., Šardzík, R., Eyers, C.E., Flitsch, S. L., Lindhorst, T. K. (2012) Dual purpose S-trityl-linkers for glycoarray fabrication on both polystyrene and gold. *Organic & Biomolecular Chemistry* **10**:8919 – 8926.
doi: 10.1039/C2OB26118A.

[10] Weissenborn, M.J., Castangia, R., Wehner, J.W., Lindhorst, T.K., Flitsch, S.L. (2012) Oxo-Ester Mediated Native Chemical Ligation on Microarrays: An Efficient and Chemoselective Coupling Methodology. *Chemical Communications* **48**:4444 – 4446.
doi: 10.1039/C2CC30844D.

[11] Fessele, C., Lindhorst, T.K. (2013) Effect of aminophenyl and aminothiahexyl α-D-glycosides of the *manno*-, *gluco*-, and *galacto*-series on type 1 fimbriae-mediated adhesion of *Escherichia coli*. *Biology* **2**:1135 – 1149.
doi: 10.3390/biology2031135.

[12] Ohlsen, K., Oelschlaeger, T.A., Hacker, J., Khan, A.S. (2009) Carbohydrate receptors of bacterial adhesins: Implications and reflections. *Topics in Current Chemistry* **288**:109 – 120.
doi: 10.1007/128_2008_10.

[13] Hartmann, M., Lindhorst, T.K. (2011) The Bacterial Lectin FimH, a Target for Drug Discovery – Carbohydrate Inhibitors of Type 1 Fimbriae-Mediated Bacterial Adhesion. *European Journal of Organic Chemistry* **2011**:3583 – 3609.
doi: 10.1002/ejoc.201100407.

[14] Ernst, B., Magnani, J.L. (2009) From carbohydrate leads to glycomimetic drugs. Nature Reviews Drug Discovery **8**:661–677.
doi: 10.1038/nrd2852

[15] Wehner, J.W., Hartmann, M., Lindhorst, T.K. (2013) Are multivalent cluster glycosides a means of controlling ligand density of glycoarrays? *Carbohydrate Research* **371**:22 – 31.
doi: 10.1016/j.carres.2013.01.023.

[16] Godula, K., Bertozzi, C.R. (2012) Density Variant Glycan Microarray for Evaluating Cross-Linking of Mucin-like Glycoconjugates by Lectins. *Journal of the American Chemical Society* **134**:15732 – 15742.
doi: 10.1021/ja302193u.

[17] Zhang, Y., Li, Q., Rodriguez, L.G., Gildersleeve, J.C. (2010) An Array-Based Method To Identify Multivalent Inhibitors. *Journal of the American Chemical Society* **132**:9653 – 9662.
doi: 10.1021/ja100608w.

[18] Love, J.C., Estroff, L.A., Kriebel, J.K., Nuzzo, R.G., Whitesides, G.M. (2005) Self-Assembled Monolayers of Thiolates on Metals as a Form of Nanotechnology. *Chemical Reviews* **105**:1103 – 1169.
doi: 10.1021/cr0300789.

[19] Kind, M., Wöll, C. (2009) Organic surfaces exposed by self-assembled organothiol monolayers: preparation, characterization, and application. *Progress in Surface Science* **84**:230 – 278.
doi: 10.1016/j.progsurf.2009.06.001.

[20] Celestin, M., Krishnan, S., Bhansali, S., Stefanakos, E., Goswami, D.Y. (2014) A review of self-assembled monolayers as potential THz frequency tunnel diodes. *Nano Research* **7**:589 – 625.
doi: 10.1007/s12274-014-0429-8..

[21] Grabosch, C., Kind, M., Gies, Y., Schweighöfer, F., Terfort, A., Lindhorst, T.K. (2013) A ‘dual click’ strategy for the fabrication of bioselective, glycosylated self-assembled monolayers as glycocalyx models. *Organic & Biomolecular Chemistry* **11**:4006 – 4015.
doi: 10.1039/C3OB40386F.

[22] Hamon, F., Djedaini-Pilard, F., Barbot, F., Len, C. (2009) Azobenzene-synthesis and carbohydrate applications. *Tetrahedron* **65**:10105 – 10123.
doi: 10.1016/j.tet.2009.08.063.

[23] Russew, M.-M., Hecht, S. (2010) Photoswitches: From Molecules to Materials. *Advanced Materials* **22**:3348 – 3360.
doi: 10.1002/adma.200904102.

[24] Kramer, R.H., Fortin, D.L., Trauner, D. (2009) New photochemical tools for controlling neuronal activity. *Current Opinion in Neurobiology* **19**:1 – 9.
doi: 10.1016/j.conb.2009.09.004.

[25] Chandrasekaran, V., Lindhorst, T.K. (2012) Sweet switches: Azobenzene glycoconjugates by click chemistry. *Chemical Communications* **48**:7519 – 7521.
doi: 10.1039/C2CC33542E.

[26] Chandrasekaran, V., Kolbe, K., Beiroth, F, Lindhorst, T.K. (2013) Synthesis and testing of the first azobenzene mannobioside as photoswitchable ligand for the bacterial lectin FimH. *Beilstein Journal of Organic Chemistry* **9**:223 – 233.
doi: 10.3762/bjoc.9.26.

[27] Chandrasekaran, V., Johannes, E., Kobarg, H., Sönnichsen, F.D., Lindhorst, T.K. (2014) Synthesis and photochemical properties of configurationally varied azobenzene glycosides. *Chemistry Open* **3**:78.
doi: 10.1002/open.201402016.

[28] Hartmann, M., Papavlassopoulos, H., Chandrasekaran, V., Grabosch, C., Beiroth, F., Lindhorst, T.K., Röhl, C. (2012) Activity and Cyctotoxicity of Five Synthetic Mannosides as Inhibitors of Bacterial Adhesion. *FEBS Letters* **586**:1459 – 1465.
doi: 10.1016/j.febslet.2012.03.059.

[29] Poloni, C., Szymański, W., Hou, L., Browne, W.R., Feringa, B.L. (2014) A Fast, Visible-Light-Sensitive Azobenzene for Bioorthogonal Ligation. *Chemistry – A European Journal* **20**:946 – 951.
doi: 10.1002/chem.201304129.

[30] Chandrasekaran, V., Jacob, H., Petersen, F., Kathirvel, K., Tuczek, F. Lindhorst, T.K. (2014) Synthesis and surface-spectroscopic characterization of photoisomerizable glyco-SAMs on Au(111). *Chemistry – A European Journal* **20**:8744 – 8752.
doi: 10.1002/chem.201402075

Discovering the Subtleties of Sugars
June 10th – 14th, 2013, Potsdam, Germany

Tools to MSn Sequence and Document the Structures of Glycan Epitopes

Hailong Zhang[1], David J. Ashline[2] and Vernon N. Reinhold[1,2,*]

[1]The Glycomics Center, University of New Hampshire, Durham, NH 03824, USA

[2]Glycan Connections, LLC, Lee, NH 03861, USA

E-Mail: *vnr@unh.edu

Received: 6th May 2014 / Published: 22nd December 2014

Abstract

Sequential disassembly (MSn) has been applied to fully characterise and document native samples containing glycan epitopes with their synthetic analogues. Both sample types were prepared by methylation, solvent phase extracted, directly infused and spatially resolved. Product ions of all samples were compiled and contrasted using management tools prepared for the fragment ion library. Each of the epitopes was further disassembled to confirm the multiple structural isomers probable within component substructures of linkage and branching. All native samples tested proved to be matched with their synthetic analogues and reasonably identical on either linear or cylindrical ion traps. Not surprisingly, spectra of mixed epitopes fragment independently, being uninfluenced by similarities. The approach has been coupled with computational tools for data handling and presentation.

Introduction

Glycan epitopes: components of function

Accumulating evidence indicates that carbohydrate glycans are participants in numerous functional roles as a consequence of their interactions with inter- and intracellular ligands. It is also clear that much of this activity is attributed to a small set of residues, often referred

This article is part of the Proceedings of the Beilstein Glyco-Bioinformatics Symposium 2013.
www.proceedings.beilstein-symposia.org

to as epitopes or glycotopes. Despite the functional importance of these epitopes, the current methods used leave numerous deficiencies. The protocols usually include lectin trapping, antibody and enzymatic assays, and multiple forms of chromatography in conjunction with MS and/or MS/MS for structural understanding. Epitopes (glycotopes) are frequently distal oligosaccharide substructures that mediate biological function through their interactions with various inter- and intracellular ligands [1]. Such glycotopes are involved in cell adhesion, blocking, signalling and increasingly are considered as disease biomarkers [2 – 6]. Examples of these include the Lewis (Le) series: Le^a, Le^x, and their sialylated analogues (SLe^a, SLe^x). These latter structures have been also considered to play pivotal roles in tumour metastasis [7]. Many recombinant biopharmaceutical drugs have glycosylation motifs that originate with host cell lines. Such products may be incompatible (antigenic) with human tissues. As one example, the *N*-glycan containing a Galα1 – 3Gal sequence (known as the α-Gal epitope) is commonly detected in murine-derived glycoproteins. Such α-Gal epitopes are highly antigenic to humans [8], and have even been observed in Chinese Hamster Ovary (CHO) cell lines [9]. Therefore, sensitive and reliable structural characterisation of these samples has significant implications in biopharmaceutical QA/QC applications.

Bringing glycomic analysis to comprehension

Unlike proteins, glycans are not direct gene products but are synthesised through a series of step-wise additions catalysed by glycosyltransferases and often modulated by the proximal environment [10 – 12]. Although it has been attempted [13], glycan structures cannot be predicted by following gene expression data. Instead, elucidation requires direct analysis of the sample itself. Moreover, glycans are often branched with numerous structural and stereo isomers. The stereo-chemical variations in each monomer coupled with the multiple sites of oligomer linkage results in an astonishing number of possible structures [14]. These intrinsic properties and problems make comprehensive oligosaccharide sequencing a fundamental as well as a challenging task.

Considerable effort has been exerted in the development of glycan characterisation methodologies. Different from DNA/RNA, glycan samples cannot be amplified. Therefore, sequencing technologies need to operate with minute quantities, often eliminating Nuclear Magnetic Resonance (NMR) as a solution. Structural analysis may be augmented by the use of specific glycosidases [15], but here specificity is frequently local, missing larger topological details of antennae. Additionally, troublesome are factors such as cost, availability, and impurities which all introduce challenges [16]. Currently, the presence of a specific glycoconjugate epitope is determined by lectin, antibody-binding, or enzymatic assay [17, 18]. While these methods can shed light on the underlying structures, it is not uncommon to find that the results from different methods may contradict each other and inevitably the resulting conclusions can be misleading. Importantly, many glycotopes have isomeric analogues; for example, the Le^a and Le^x antigens are structural isomers of galactose, fucose, and *N*-acetyl glucosamine (GlcNAc), with the linkage positions of fucose and galactose reversed on GlcNAc. Often, separation techniques cannot resolve these structural isomers effectively. While liquid chromatography in conjunction with MS and MS/MS, plus glycosidase treatment, is common. Such described data often includes considerable inference and intuition of know biological systems, failing to provide a clear and complete picture [19]. In summary, the current strategies for glycan characterisations are found suboptimal [20], and MS and MS/MS applications are incomplete.

Sequential mass spectrometry (MSn): a comprehensive approach

Repetitive steps of disassembly (MSn) in an ion trap mass spectrometer provide ion compositions and spectral products that match the stereo and structural isomers of standards. Such products, along with metal binding, and know fragment pathways associated with activation provide insight for reassembly [21 – 23]. This approach has been used successfully to characterise unique and unusual structural features [24], as well as understanding to isomeric mixtures [25 – 28]. Evidence has shown that MSn spectra of glycotopes are highly reproducible and spectra of standards have been used for documentation. Library matching and *de novo* bioinformatics tools have been developed to handle these data sets [29 – 32].

Methods and Results

Analytical methods: sample preparation and mass spectrometry

In this study various standard biologicals and synthetic epitopes were prepared and processed providing fragment ion spectral products for library documentation. The workup requires methylation, solid phase extraction, and MSn analysis. Precursor ions of synthetic and natural epitopes were isolated, and each disassembled and contrasted for identity. The resulting spectra and pathways of disassembly provided opportunities to document exacting and comprehensive structural detail from the searchable fragment library.

The samples studied include: (1) *N*-linked glycans from human plasma; (2) *O*-linked glycans from human colon tissue; (3) *N*-linked glycans from human breast cancer cell line MCF-7; (4) Lacto-*N*-difucohexaose I (LNDFH I; purified from human milk), a standard containing Lewis B; (5) CFG Te118 (Fucα1-2Galβ1-4[Fucα1-3]GlcNAcβ-CH2CH2N3), a synthetic standard containing Lewis Y; (6) CFG Te140 (Neu5Acα2-3(Galβ1-4[Fucα1-3]GlcNAcβ1-3)2-CH2CH2N3), a synthetic standard containing Sialyl-Lewis X.

For human plasma samples, blood was collected in a Becton Dickinson vacutainer (East Rutherford, NJ, USA) containing sodium citrate as an anticoagulant. Plasma was separated by centrifugation and passed through a Protein A/G column ThermoFisher/Pierce (Rockford, IL, USA) to obtain IgG-depleted plasma and IgG fractions. *N*-linked glycans from IgG-depleted human plasma were released with *N*-glycanase (Prozyme).

All *N*-linked glycans were derivatised with 2-AA via reductive amination. These samples were separated by HPLC using a HILIC phase. Collected fractions were dried and permethylated. *O*-linked glycans were released via β-elimination using alkaline borohydride [33]. Samples were purified via cation exchange and porous graphitised carbon solid phase extraction prior to permethylation.

Permethylation was carried out in spin columns Harvard Apparatus (Holliston, MA, USA) as described [34]. Sodium hydroxide beads and iodomethane were purchased from Sigma. Purification of permethylated oligosaccharides was performed by liquid–liquid extraction with dichloromethane and 0.5 M aqueous sodium chloride.

LNDFH I (GKAD-02010) was acquired from Prozyme (Hayward, CA, USA). Other synthetic glycan standards were graciously supplied by the Consortium for Functional Glycomics (CFG). All standard materials were permethylated, extracted and used directly.

The methylated oligosaccharides were dissolved in 1:1 methanol/water. The samples were loaded onto a Triversa Nanomate Advion (Ithaca, NY, USA) mounted onto an ion trap mass spectrometer (LTQ, ThermoFisher). Activation Q and activation time were left at the default values, 0.250 and 30 ms, respectively. Collision energy was set to 35% for all CID spectra. The scan rate was set to "Enhanced". Data were acquired and are displayed in profile mode. MSn peaks were selected manually. In general, the precursor ion mass window was set to capture the full isotopic envelope. This was done to obtain complete isotope clusters of fragment ions at higher stages of MSn, and to simplify fragment ion charge state determination. For each data file, at least one scan was obtained of the isolated precursor ion with the collision energy set to 0, to record the precursor isotope envelope. The signal was averaged for a variable number of scans, with the times indicated in each spectrum. The microscan count, AGC target value, and maximum injection times were varied, depending on the signal intensity. All ions are sodium adducts.

Bioinformatics tools: spectral data handling and scoring

The MSn spectral matching based glycotope identification strategy introduces a unique component to data processing. For instance, the MSn spectra acquired are in a proprietary native binary file format, which cannot be accessed and processed directly by the existing mass spectral searching engines. Although some peak list extraction utility tools do exist, these tools are usually designed for proteomics data: only MS and MS/MS data are supported; and the tools cannot average multiple scans/microscans contained in MSn spectra. To improve spectra quality, the MSn based approach often requires multiple micro-scans to be collected and averaged for each target precursor. An automated MSn experiment on the Thermo LTQ can easily result in hundreds of microscans from multiple precursors. None of the existing tools are designed for this type of data handling. The desired software tool must be able to automatically (1) extract MSn spectra peak lists from native data files;

(2) preserve the MSn precursor relationships; (3) organise spectra by precursors; and (4) average microscans for each MSn spectrum. We have developed a bioinformatics package FragLib Tool Kit to fulfil these requirements.

FragLib Tool Kit is a collection of software tools designed for building MSn spectra libraries and facilitating data handling and management tasks. Through the Application Programming Interface (API) provided by Thermo Xcalibur Development Kit (XDK) COM library, FragLib Tool Kit is capable of accessing native Thermo Xcalibur raw spectral files directly. The tool can automate the tasks of MSn spectra archiving. Instead of simply compiling peak lists, FragLib Tool Kit can capture the essential information of raw data files, and convert the data into NIST MS tool which is compatible with mass spectral libraries, which have much smaller disk footprint. For instance, one recent library archive contains ~90,000 MSn spectra. The original data files require ~100 GB disk space and cannot be searched easily. The size of the searchable archive library generated by FragLib Tool Kit is only ~30 MB, which can be carried easily with a USB flash drive. FragLib Tool Kit is developed in C#. Thermo Xcalibur XDK and Microsoft .Net framework are required to install and run the tools properly. The highlights of functionalities offered by FragLib Tool Kit include: (1) Archive Xcalibur raw spectral files directly (in both interactive and daemon modes): The predefined regular archiving tasks can be automated under the daemon mode using the tool built-in archiving scheduler; (2) Search/Filter spectral archive records using multiple query constrains; (3) Support attaching structural annotation to archive records; (4) Convert archives into NIST MSP and other popular MS library formats; (5) Build mass spectral libraries in NIST format; (6) Compute spectrum matching scores to quantify the similarity between given input spectra.

Spectrum matching scores were generated using the following formula:

$$\text{Similarity score} = \cos\theta = \frac{\mathrm{u}\cdot\mathrm{v}}{\sqrt{\mathrm{u}\cdot\mathrm{u}}\sqrt{\mathrm{v}\cdot\mathrm{v}}}$$

Where "·" indicates the dot product, u and v are the aligned mass spectral intensity vectors of the two input spectra, with the intensity values ranging from 0 to 100. Using u to denote the vector of the observed spectrum and v to denote the vector of the standard, the most abundant peak has an intensity of 100 and all other peaks are normalised accordingly. The resulting similarity score is a numeric measurement of the spectral similarity, between 0 and 1, where 1 indicates that the spectra are identical and 0 indicates that no similarity exists. For our purposes, a similarity score greater than 0.850 is strongly indicative of a match using the described scoring algorithm.

A similarity score of two given mass spectra can be visualised as the cosine of the angle between the intensity vectors of the two spectra. Figure 1 depicts the visualisation of similarity score (Figure 1c) using two simplified input spectra: spectrum A (Figure 1a) and spectrum B (Figure 1b). When the two spectra match perfectly, the two intensity vectors overlap: the angle between the two vectors is 0, so the similarity score is 1 indicating a perfect spectra match. On the other hand, while the difference between two spectra increases, the angle between the vectors increases, and the similarity score decreases indicating a greater difference between the two spectra.

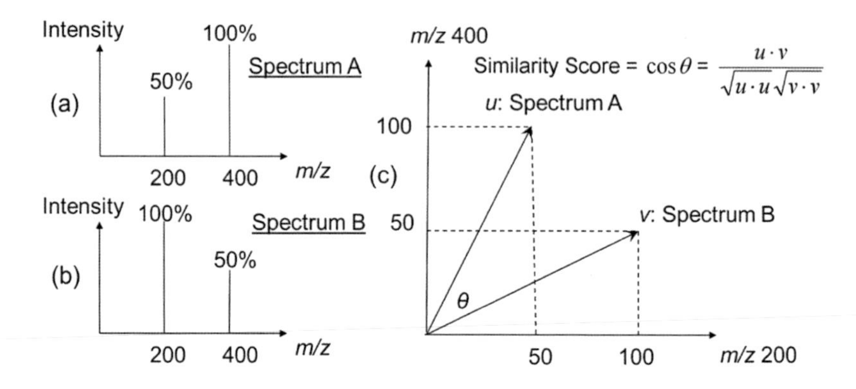

Figure 1. Visualisation of Similarity Score **(c)** using two simplified mass spectra: Spectrum A **(a)** and Spectrum B **(b)**. A Similarity Score of two given mass spectra can be visualised as the cosine of the angle between the intensity vectors of the two input mass spectra.

Application example 1: Sialylated Lewis X, B-Type Tetrasaccharide (m/z 1021) and B/Y-Type Trisaccharide (m/z 646)

Figure 2 shows a comparison of Sialyl-Lewis X substructure fragments isolated from the synthetic standard, CFG Te140, and one human plasma *N*-linked glycan. Typically, at the MS/MS stage, both *m/z* 646 and 1021 fragments are detectable; sometimes the *m/z* 1021 fragment ion is very weak or not detectable. CID of the *m/z* 1021 ion produces the spectra shown in Figure 2a, b, and c, isolated from CFG Te140 (Figure 2a), a human plasma *N*-glycan (Figure 2b), and a normal human colon tissue *O*-glycan (Figure 2c), respectively. These spectra are relatively simple because of the liability of the sialic acid linkage favouring formation of the *m/z* 646 fragment. Further, CID of the *m/z* 646 ion provides a much more informative spectrum (Figures 2 d, 2e, and 2f). All three samples exhibit very similar spectra, both in terms of fragment masses and the overall intensity pattern, despite the differences in sample and precursor ion structures. Figure 3 shows putative fragment assignments for this structure. As with Lewis X, the 3,5A-type cleavage across the *N*-acetyl-glucosamine residue positions the galactose residue at the 4-position rather than the 3-position. This provides additional *de novo* evidence of the SLe^x structure rather than SLe^a. Further comparison could be made between these same mass fragments generated from a sialylated Lewis A standard. Reliable mass spectrometric distinction of SLe^x and SLe^a would have significant potential utility [33]. Unfortunately, we are not aware of any suitable standard that is available. The SLe^a tetrasaccharide is commercially available, but is unsuitable for this analysis, as it would not produce the correct B-type fragment. One could use the tetrasaccharide to prepare a glycoside (other than methyl) prior to permethylation. In our prior experience, although such structures produce the same mass fragments, they produce different intensity patterns than is obtained from cleavage of glycosidic bonds. The reasons behind this are not well understood; however, a suitable standard typically must have at least two monosaccharide residues between the epitope of interest and any aglycone functional group to be used in this way.

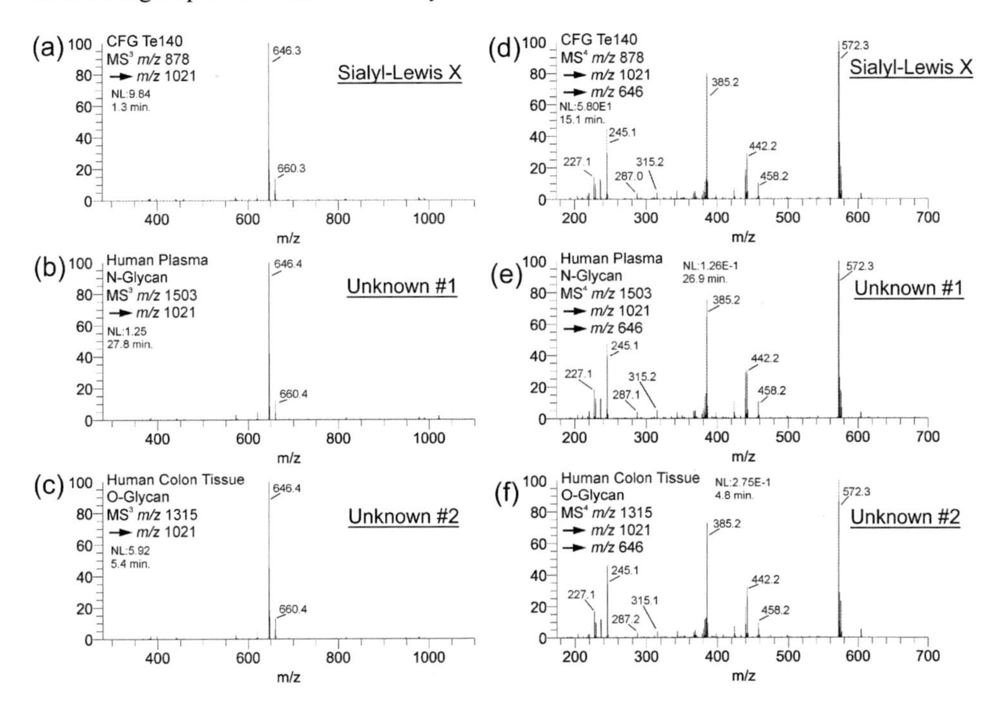

Figure 2. Comparison of B-type sialyl-Lewis X tetrasaccharide ion and B/Y-type trisaccharide ion spectra. Spectra **(a)**, **(b)**, and **(c)** show the tetrasaccharide spectra of CFG Te140 (sialyl-Lewis X standard), human plasma *N*-glycan, and normal human colon tissue *O*-glycan, respectively. Spectra **(d)**, **(e)**, and **(f)** show the trisaccharide spectra of CFG Te140 (sialyl-Lewis X standard), human plasma *N*-glycan, and normal human colon tissue, respectively. At the tetrasaccharide level, the spectra are dominated by NeuAc loss; at the trisaccharide level (*m/z* 646), the fragmentation is much more informative.

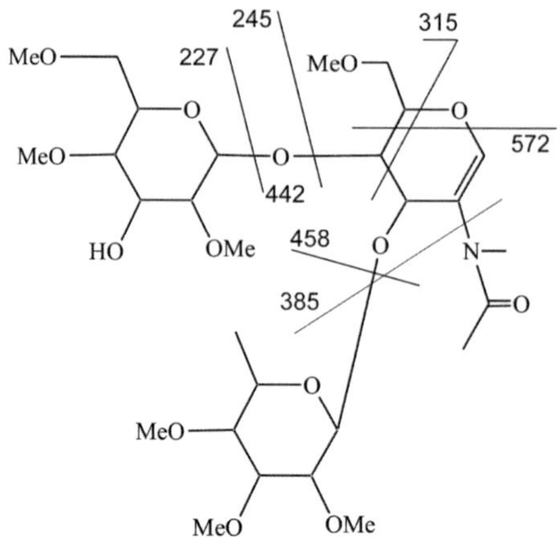

Figure 3. Structure and fragmentation of sialyl-Lewis X fragment ion m/z 646.

Application example 2: Lewis Y, Lewis B, B-Type Tetrasaccharide (m/z 834), and B/Y Trisaccharide (m/z 646)

Lewis Y is a tumour-associated carbohydrate antigen [35 – 37]. Studies have suggested that it plays a role in cell cycle perturbation [37], and it has been selected as the target of vaccine development [38]. Lewis B is also a difucosylated Hex-HexNAc, but built upon a type 1 lactosamine. Figure 4a and Figure 4 d show the MS 3 and MS 4 spectra of the tetrasaccharide B ion *m/z* 834 and trisaccharide B/Y ion *m/z* 646, respectively, for a Lewis B standard (LNDFH I). Figure 4b and Figure 4e show the same spectra isolated from a Lewis Y standard (CFG Te118). At the MS 3 level (Figure 4a and Figure 4b), the spectra show subtle intensity pattern differences, but they do not clearly distinguish these two structures. At the trisaccharide level (Figure 4 d and Figure 4e), however, the differences are much more pronounced and easily serve to distinguish the two structures. Figure 4c and Figure 4f show the fragmentation spectra with the same *m/z* precursor obtained from an *N*-glycan antenna (precursor *m/z* 16913+, composition Hex10HexNAc9dHex4) isolated from MCF-7 cultured cells. The MS 3 spectra are virtually identical to those of the Lewis Y standard (similarity score 0.952) despite the nearly three orders of magnitude greater intensity for the standard spectra (NL 1.03E4 vs. NL 4.41E1); the MS 4 spectra shown are also identical (similarity score 0.996), despite four orders of magnitude difference in intensity (NL 9.24E2 vs. NL 6.74E-2).

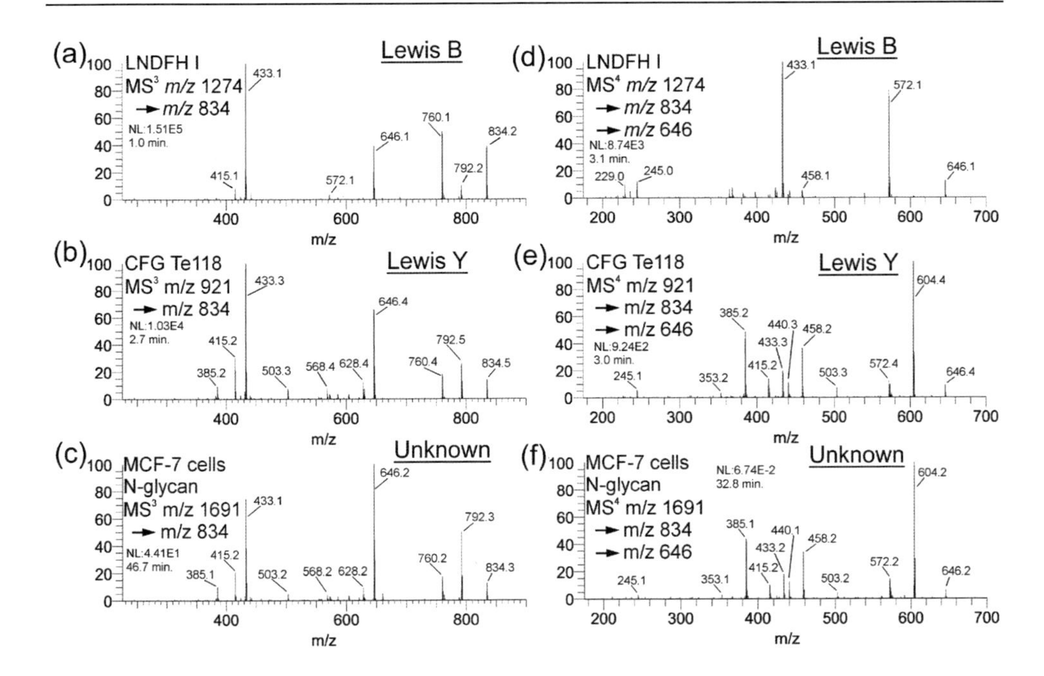

Figure 4. Comparison of spectra of B-type Lewis B and Lewis Y tetrasaccharide and B/Y-type trisaccharide ions. Standard spectra **(a) - (d)** and **(b) – (e)** were obtained from lacto-*N*-difucohexaose I (Lewis B) and CFG Te118 (Lewis Y), respectively. Spectra **(c) - (f)** represent the analogous spectra isolated from an *N*-glycan (Hex10HexNAc9dHex4, triply-charged *m/z* 1691^{3+}) of MCF-7 cultured cells.

CONCLUSION

Antibodies and lectins are widely accepted for performing specific analytical roles in glycomic studies; however, mass spectrometry offers a broader range of supplemental information, often with extending insight; a neutral loss, positive/negative ion extraction, alternative adducting ions and their compositions. Such considerations all contribute component understanding at minimal cost and sample loss. In addition, MSn can resolve mixtures even in the presence of multiple isomers [24, 25, 27]. While *de novo* structural assignments rely on fragment composition masses as the primary information, spectral intensity can be most specific when resolving subtle structural differences and here, synthetic standards are excellent ways of confirmation. Sample impurities can be worked with by considering alternative derivatisation strategies coupled with selective solvent phase extraction. If ionisation of the parent or precursor can be achieved MS 2 ion abundances are not influenced in product ion patterns. Structural problems of this type require comparison of spectra to standard materials with known structure and high purity to provide relevant benchmarks. A requirement for choosing suitable standard materials is that they must be larger than the substructure of interest to provide for fragment generation inside the ion trap mass spectrometer. Ideally, standards are larger oligosaccharides containing the epitope of interest. Synthetic glycoconjugates with an aglycone, such as the azido linkers found on the CFG standards, are also useful.

Inevitable to glycomic methodological discussions are questions of sensitivity and required sample amounts. While typical sequential mass spectrometric analyses may require somewhat greater sample amounts than fluorescence chromatography or online LC-MS, the level of information acquired versus the level of structural detail needed to answer a particular question also should be considered. It has been frequently acknowledged that glycans offer a tremendous challenge because of their structural complexity and the potential for multiple isomeric forms, especially among larger oligosaccharides [14]. Although a multitude of methods are available that require very little sample, these also provide comparatively less structural information. While sample quantities may limit the depth of analysis, the informational limitations of the chosen methodologies should not be ignored, and structures that are assumed from mass composition only should also be acknowledged as such.

The possibility of isomerism in large complex oligosaccharides remains a frequently ignored or oversimplified analytical problem, owing to the inability of many common methods to approach this issue. In short, the comparison of methods primarily in terms of sensitivity disregards the relative amounts of information obtainable using the chosen strategies. Having mixtures of isomers in biological samples should be considered a routine occurrence, and the absence of isomers the unusual event. Having the library of epitope MSn finger-printing spectra, of pure materials, can make mixture spectra more easily interpretable by realising that they can be considered as superpositions of the relevant standard spectra. Thus, spectra generated from biological materials can be matched against multiple standard spectra, of the same precursor mass, to ascertain the likelihood that the unknown spectrum is a superposition of pure spectra, and therefore a mixture of isomers.

The increasing availability of interesting synthetic oligosaccharides has made the generation of oligosaccharide MSn libraries a viable goal and will vastly extend the utility of this technique to any analysts with the appropriate sample preparation and instrumentation. Moreover, Thermo has made the API to an instrument control COM object library publicly accessible to provide a programming interface to allow for custom software codes to control Thermo IT instruments. Using the COM library API, one can control the instrument at a high level without needing to know low level hardware control details. By integrating our MSn library with the Thermo Instrument Control COM, we would be able to direct the instrument to perform data-dependent MSn experiments using prior knowledge accumulated in an MSn library. At present, we are exploring and evaluating this possibility for future development of our MSn bioinformatics tools.

ACKNOWLEDGEMENTS

The authors thank James M. Paulson (Scripps Research Institute) and the Consortium for Functional Glycomics for the synthetic standards, Robert Sackstein and Cristina I. Silvescu (Brigham and Women's Hospital) for the human colon tissue, Dipak K. Banerjee (University of Puerto Rico) for the MCF-7 cultured cells.

This work was supported in part by an NIH Program of Excellence in Glycosciences grant (P01 HL107146, PI: Dr. Robert Sackstein) and Glycan Connections, LLC, Lee, NH 03861, USA.

REFERENCES

[1] Cummings, R.D. (2009) The repertoire of glycan determinants in the human glycome. *Molecular BioSystems* **5**:1087 – 104.
doi: 10.1039/B907931A.

[2] Burdick, M.M., McCaffery, J.M., Kim, Y.S., Bochner, B.S., Konstantopoulos, K. (2003) Colon carcinoma cell glycolipids, integrins, and other glycoproteins mediate adhesion to HUVECs under flow. *American Journal of Physiology Cell Physiology* **284**:C977 – 987.
doi: 10.1152/ajpcell.00423.2002.

[3] Kannagi, R. (2007) Carbohydrate antigen sialyl Lewis a – Its pathophysiological significance and induction mechanism in cancer progression. *Chang Gung Medical Journal* **30**:189 – 209.

[4] Harder, J., Kummer, O., Olschewski, M., Otto, F., Blum, H.E., Opitz, O. (2007) Prognostic relevance of carbohydrate antigen 19 – 9 levels in patients with advanced biliary tract cancer. *Cancer Epidemiology, Biomarkers & Prevention* **16**:2097 – 2100.
doi: 10.1158/1055-9965.EPI-07-0155.

[5] Aoyama, H., Tobaru, Y., Tomiyama, R., Maeda, K., Kishimoto, K., Hirata, T., Hokama, A., Kinjo, F., Fujita, J. (2007) Elevated carbohydrate antigen 19 – 9 caused by early colon cancer treated with endoscopic mucosal resection. *Digestive Diseases and Sciences* **52**:2221 – 2214.
doi: 10.1007/s10620-006-9247-5.

[6] Arata-Kawai, H., Singer, M.S., Bistrup, A., Zante, A., Wang, Y.Q., Ito, Y., Bao, X., Hemmerich, S., Fukuda, M., Rosen, S.D. (2011) Functional contributions of *N*- and *O*-glycans to L-selectin ligands in murine and human lymphoid organs. *The American Journal of Pathology* **178**:423 – 433.
doi: 10.1016/j.ajpath.2010.11.009.

[7] Jacobs, P.P., Sackstein, R. (2011) CD44 and HCELL: Preventing hematogenous metastasis at step 1. *FEBS Letters* **585**(20):3148 – 3158.
doi: 10.1016/j.febslet.2011.07.039.

[8] Macher, B.A., Galili, U. (2008) The Gal alpha 1,3Gal beta 1,4GlcNAc-R (α-Gal) epitope: a carbohydrate of unique evolution and clinical relevance. *Biochimica et Biophysica Acta (BBA) General Subjects* **1780**:75 – 88.
doi: 10.1016/j.bbagen.2007.11.003.

[9] Bosques, C.J., Collins, B.E., Meador, J.W. 3rd, Sarvaiya, H., Murphy, J.L., Dellorusso, G., Bulik, D.A., Hsu, I.H., Washburn, N., Sipsey, S.F., Myette, J.R., Raman, R., Shriver, Z., Sasisekharan, R., Venkataraman, G. (2010) Chinese hamster ovary cells can produce galactose-alpha-1,3-galactose antigens on proteins. *Nature Biotechnology* **28**:1153 – 1156.
doi: 10.1038/nbt1110-1153.

[10] Schachter, H. (2000) The joys of HexNAc. The synthesis and function of *N*- and *O*-glycan branches. *Glycoconjugate Journal* **17**:465 – 83.
doi: 10.1023/A:1011010206774

[11] Kornfeld, R., Kornfeld, S. (1985) Assembly of asparagine-linked oligosaccharides. *Annual Review of Biochemistry* **54**:631 – 664.
doi: 10.1146/annurev.bi.54.070185.003215.

[12] Hanisch, F.G. (2001) *O*-glycosylation of the mucin type. *Biological Chemistry* **382**:143 – 149.
doi: 10.1515/BC.2001.022.

[13] Kawano, S., Hashimoto, K., Miyama, T., Goto, S., Kanehisa, M. (2005) Prediction of glycan structures from gene expression data based on glycosyltransferase reactions. *Bioinformatics* **21**:3976 – 3982.
doi: 10.1093/bioinformatics/bti666.

[14] Laine, R.A. (1994) A calculation of all possible oligosaccharide isomers both branched and linear yields 1.05 × 1012 structures for a reducing hexasaccharide: the Isomer Barrier to development of single-method saccharide sequencing or synthesis systems. *Glycobiology* **4**:759 – 767.
doi: 10.1093/glycob/4.6.759.

[15] Kuster, B., Naven, T.J., Harvey, D.J. (1996) Rapid approach for sequencing neutral oligosaccharides by exoglycosidase digestion and matrix-assisted laser desorption/ionization time-of-flight mass spectrometry. *Journal of Mass Spectrometry* **31**: 1131 – 1140.
doi: 10.1002/(sici)1096-9888(199610)31:10<1131::aid-jms401>3.0.co;2-r.

[16] Geyer, H., Geyer, R. (2006) Strategies for analysis of glycoprotein glycosylation. *Biochimica et Biophysica Acta (BBA) – Proteins and Proteomics* **1764**:1853 – 1869. doi: 10.1016/j.bbapap.2006.10.007.

[17] Croce, M.V., Sálice, V.C., Lacunza, E., Segal-Eiras, A. (2005) Alpha 1-acid glycoprotein (AGP): a possible carrier of sialyl lewis X (slewis X) antigen in colorectal carcinoma. *Histology and Histopathology* **20**:91 – 97.

[18] Li, C., Zolotarevsky, E., Thompson, I., Anderson, M.A., Simeone, D.M., Casper, J.M., Mullenix, M.C., Lubman, D.M. (2011) A multiplexed bead assay for profiling glycosylation patterns on serum protein biomarkers of pancreatic cancer. *Electrophoresis* **32**:2028 – 2035.
doi: 10.1002/elps.201000693.

[19] Anumula, K.R. (2006) Advances in fluorescence derivatization methods for high-performance liquid chromatographic analysis of glycoprotein carbohydrates. *Analytical Biochemistry* **350**:1 – 23.
doi: 10.1016/j.ab.2005.09.037.

[20] Parry, S., Ledger, V., Tissot, B., Haslam, S.M., Scott, J., Morris, H.R., Dell, A. (2007) Integrated mass spectrometric strategy for characterizing the glycans from glycosphingolipids and glycoproteins: direct identification of sialyl Le(x) in mice. *Glycobiology* **17**:646 – 654.
doi: 10.1093/glycob/cwm024.

[21] Reinhold, V.N., Sheeley, D.M. (1998) Detailed characterization of carbohydrate linkage and sequence in an ion trap mass spectrometer: glycosphingolipids. *Analytical Biochemistry* **259**:28 – 33.
doi: 10.1006/abio.1998.2619.

[22] Sheeley, D.M., Reinhold, V.N. (1998) Structural characterization of carbohydrate sequence, linkage, and branching in a quadrupole Ion trap mass spectrometer: neutral oligosaccharides and *N*-linked glycans. *Analytical Chemistry* **70**:3053 – 3059.
doi: 10.1021/ac9713058.

[23] Ashline, D.. Singh, S.. Hanneman, A.. Reinhold, V. (2005) Congruent strategies for carbohydrate sequencing. 1. Mining structural details by MSn. *Analytical Chemistry* **77**:6250 – 6262.
doi: 10.1021/ac050724z.

[24] Hanneman, A.J., Rosa, J.C., Ashline, D., Reinhold, V.N. (2006) Isomer and glycomer complexities of core GlcNAcs in *Caenorhabditis elegans*. *Glycobiology* **16**:874 – 890.
doi: 10.1093/glycob/cwl011.

[25] Ashline, D.J., Lapadula, A.J., Liu, Y.H., Lin, M., Grace, M., Pramanik, B., Reinhold, V.N. (2007) Carbohydrate structural isomers analyzed by sequential mass spectrometry. *Analytical Chemistry* **79**:3830 – 3842.
doi: 10.1021/ac062383a.

[26] Jiao, J., Zhang, H., Reinhold, V.N. (2011) High Performance IT-MS Sequencing of Glycans (Spatial Resolution of Ovalbumin Isomers). *International Journal of* Mass *Spectrometry* **303**:109 – 117.
doi: 10.1016/j.ijms.2011.01.016.

[27] Prien, J.M., Ashline, D.J., Lapadula, A.J., Zhang, H., Reinhold, V.N. (2009) The high mannose glycans from bovine ribonuclease B isomer characterization by ion trap MS. *Journal of the American Society for Mass Spectrometry* **20**:539 – 556.
doi: 10.1016/j.jasms.2008.11.012.

[28] Prien, J.M., Huysentruyt, L.C., Ashline, D.J., Lapadula, A.J., Seyfried, T.N., Reinhold, V.N. (2008) Differentiating *N*-linked glycan structural isomers in metastatic and nonmetastatic tumor cells using sequential mass spectrometry. *Glycobiology* **18**:353 – 366.
doi: 10.1093/glycob/cwn010.

[29] Zhang, H., Singh, S., Reinhold, V.N. (2005) Congruent strategies for carbohydrate sequencing. 2. FragLib: an MSn spectral library. *Analytical Chemistry* **77**: 6263 – 6270.
doi: 10.1021/ac050725r.

[30] Lapadula, A.J., Hatcher, P.J., Hanneman, A.J., Ashline, D.J., Zhang, H., Reinhold, V.N. (2005) Congruent strategies for carbohydrate sequencing. 3. OSCAR: an algorithm for assigning oligosaccharide topology from MSn data. *Analytical Chemistry* **77**:6271 – 6279.
doi: 10.1021/ac050726j.

[31] Ashline, D.J., Hanneman, A.J., Zhang, H., Reinhold, V.N. (2014) Structural Documentation of Glycan Epitopes: Sequential Mass Spectrometry and Spectral Matching. *Journal of the American Society for Mass Spectrometry* **25**(3):444 – 453.
doi: 10.1007/s13361-013-0776-9.

[32] Reinhold, V., Zhang, H., Hanneman, A., Ashline, D. (2013) Toward a platform for comprehensive glycan sequencing. *Molecular & Cellular Proteomics* **12**:866 – 873.
doi: 10.1074/mcp.R112.026823.

[33] Carlson, D.M. (1966) Oligosaccharides Isolated from Pig Submaxillary Mucin. *The Journal of Biological Chemistry* **241**:2984 – 2986.

[34] Kang, P., Mechref, Y., Klouckova, I., Novotny, M.V. (2005) Solid-phase permethylation of glycans for mass spectrometric analysis. *Rapid Communication in Mass Spectrometry* **19**:3421 – 3428.
doi: 10.1002/rcm.2210.

[35] Madjd, Z., Parsons, T., Watson, N.F., Spendlove, I., Ellis, I., Durrant, L.G. (2005) High expression of Lewis y/b antigens is associated with decreased survival in lymph node negative breast carcinomas. *Breast Cancer Research* **7**:R780 –R787.
doi: 10.1186/bcr1305.

[36] Nudelman, E., Levery, S.B., Kaizu, T., Hakomori, S. (1986) Novel fucolipids of human adenocarcinoma: characterization of the major Ley antigen of human adenocarcinoma as trifucosylnonaosyl Ley glycolipid (III3FucV3FucVI2FucnLc6). *The Journal of Biological Chemistry* **261**:11247 – 11253.

[37] Liu, D., Liu, J., Lin, B., Liu, S., Hou, R., Hao, Y., Liu, Q., Zhang, S., Iwamori, M. (2012) Lewis-y regulate cell cycle related factors in ovarian carcinoma cell RMG-I in vitro via ERK and Akt signaling pathways. *International Journal of Molecular Sciences* **13**:828 – 839.
doi: 10.3390/ijms13010828.

[38] Heimburg-Molinaro, J., Lum, M., Vijay, G., Jain, M., Almogren, A., Rittenhouse-Olson, K. (2011) Cancer vaccines and carbohydrate epitopes. *Vaccine* **29:** 8802 – 8826.
doi: 10.1016/j.vaccine.2011.09.009.

Discovering the Subtleties of Sugars
June 10th – 14th, 2013, Potsdam, Germany

Isolation and Purification of Glycans from Natural Sources for Positive Identification

Milos V. Novotny[1,*] **and William R. Alley, Jr.**[1,2]

[1]Department of Chemistry, Indiana University, Bloomington, Indiana 47405, USA

[2]Department of Chemistry, University of Texas at San Antonio, San Antonio, TX 78249, USA

E-Mail: *novotny@indiana.edu

Received: 11th December 2013 / Published: 22nd December 2014

Abstract

The great structural diversity of glycans demands powerful analytical methodologies, such as different combinations of capillary separations with mass spectrometry (MS), to identify the correct structures involved in key glycan interactions of biological importance. Precise structural assignments, in turn, necessitate the availability of pure authentic glycans as absolute analytical standards. It is particularly evident with the cases of glycan isomerism, which are seemingly involved in the search for glyco-biomarkers of human diseases. While novel synthetic approaches are being developed toward the acquisition of new glycan standards, it is still prudent, feasible, and profitable to consider the isolation of pure glycans from some hitherto unexplored natural sources. It is demonstrated here that recycling high-performance liquid chromatography (R-HPLC) can accomplish isolation of isomeric glycans to be used as analytical standards or valuable reagents in the fabrication of glycan arrays for biomarker discovery.

This article is part of the Proceedings of the Beilstein Glyco-Bioinformatics Symposium 2013.
www.proceedings.beilstein-symposia.org

Introduction

Among the challenges of modern glycoscience is still a limited understanding of the structure–function relationship of different glycoconjugate molecules. While glycans are highly ubiquitous components of the biomolecules facilitating molecular recognition on cellular surfaces and in different levels of the immune system's function, the structural intricacies of these biologically important processes remain somewhat unclear, if not overwhelming. Glycosylation is a very complex process, often resulting in the molecules or arrays of molecules with very intricate structural forms and a very high information content [1 – 3]. Yet, we currently lack the adequate procedures to: (a) chemically resolve the structural complexity of individual glycans or measure their quantitative proportions; and (b) decipher the real biological meaning of glycosylation patterns, as based on the instrumentally measured and interpreted data. While the means to measure very small amounts of glycans and glycopeptides in biological materials have advanced remarkably during the last several years [for recent reviews, see references 4 – 6], structural elucidation and authentication of glycoconjugates have been hampered by the limited availability of glycans and glycopeptides as standard compounds that the scientific community could use in research investigations. Recently, this has been discussed in the U.S. National Research Council report [7], which calls for a concerted effort by synthetic carbohydrate chemists and analytical scientists, among others, to provide and authenticate important glycoconjugate molecules. The availability of these compounds in the future is likely to impact clinical diagnostic procedures and to improve our understanding of carbohydrate–protein and carbohydrate–carbohydrate interactions, in general.

Among the most promising approaches to probe the glycoconjugate interactions with other biologically relevant molecules have been glycan arrays [8 – 11], in which various glycan molecules are chemically immobilised to appropriate surfaces as "microdots" on a chip, thus representing diverse structures and their amounts. The glycan arrays have played a major role in the discoveries of antiglycan antibodies, galectins with different glycan binding specificities, and siglecs. Although up to several hundreds of synthetic glycans have now been available to construct glycan arrays in different laboratories [12], many more will need to become available to fill the "gaps" representing the total glycomes of different species, which are conservatively estimated to exist in up to at least several thousand structures in mammals [3]. Recent advances in the chemical and enzymatic synthesis [13 – 15] will be undoubtedly crucial in this regards, while additional efforts, as demonstrated below, could involve a yet different route to pure glycan availability, specifically, to their isolation from readily available biological materials. Appropriate bulk fractionation schemes and recovery of properly tagged glycans at high purity in the microgram-to-milligram scale need to be developed for this task.

Toward Disease Biomarker Discoveries: Recent Methodological Developments

Unusual types of protein glycosylation connected with human diseases have been the subject of investigations for several decades. However, only in recent times has it been possible to link aberrant glycan structures to pathological conditions such as congenital disorders, cardiovascular diseases, and malignancy [16, 17]. Methodological advances in measuring different oligosaccharides and their analytical profiles (quantitative glycan profiling at high sensitivity) have been of substantial value in the recognition of definitive glycan structures, with mass spectrometry (MS) and its various tandem techniques being recognised as the leading technology [for recent reviews, see references 4 – 6] in the biomarker discovery field. For the last several years, our principal motivation to develop sensitive bioanalytical instrumentation and glycomic profiling methodologies has been to explore the differences between glycosylation in healthy and in cancer conditions. While distinguishing other inflammatory diseases and cancer is currently somewhat problematic [18], glycomic profiling shows a distinct promise for the future of diagnostic and prognostic measurements [17, 19].

In cancer-related investigations through glycomics and bioinformatics, common physiological fluids (blood serum and plasma, cysts, ascitic fluids, saliva, cerebrospinal fluid, etc.) can nominally be analysed, but additionally, tumour tissues and cell lines may also provide convenient samples. While MS is uniquely sensitive to detect different glycans in these biological materials, this powerful method alone cannot determine unequivocally structural features of certain carbohydrates such as the abundance of structural isomers (glycans with the same molecular mass, but different positions or linkages of a substituted monosaccharide). Nonetheless, glycomic MS profiles can be highly informative in terms of different glycosylation patterns recorded for various cancer cohort groups [20 – 22], where the profiles obtained at high sensitivity (Figure 1) can be recorded and statistically evaluated for the patient cohorts numbering typically between 20 to 50 individuals. Various detected and quantitatively measured glycans can further be statistically evaluated for different cohort groups and judged by the usual clinical criteria (ANOVA, ROC values, etc.) whether or not they are worthy further investigation as biomarker candidates.

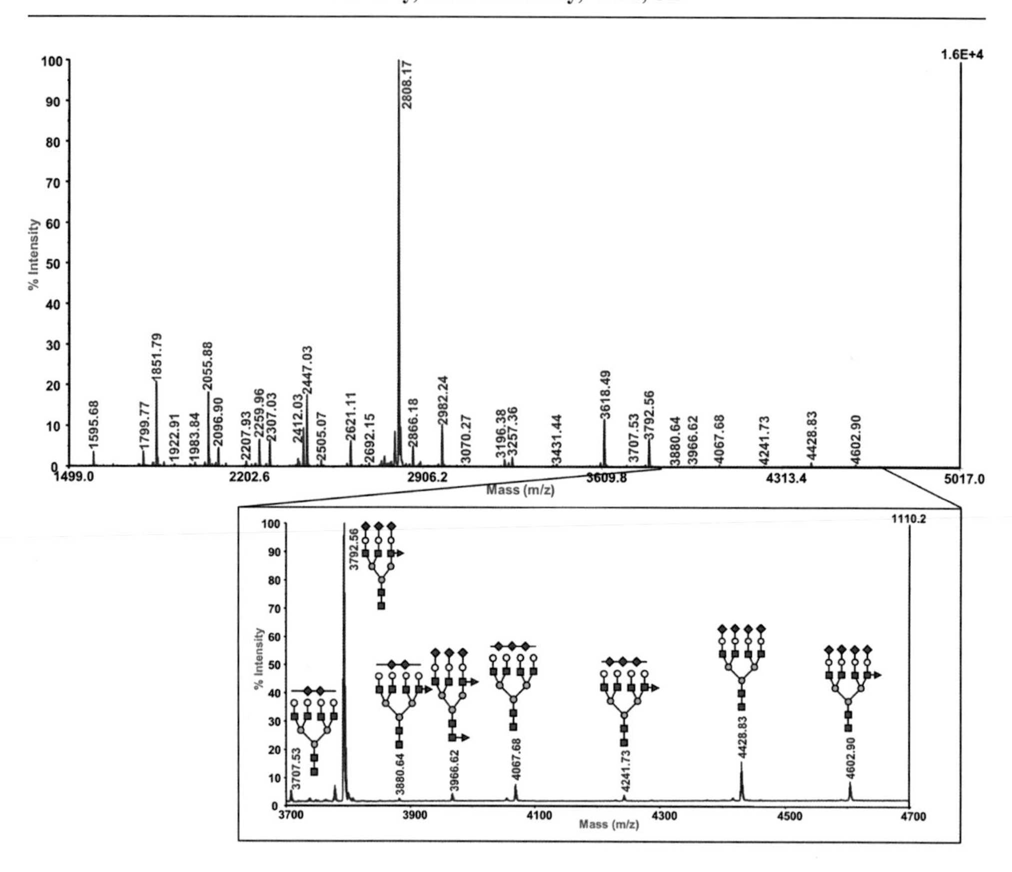

Figure 1. Example of an MS oligosaccharide profile derived from a blood serum sample measured for an ovarian cancer patient. The series of high-mass glycans (insert) are diagnostically important. Reprinted with permission from reference [21]. Copyright 2012 American Chemical Society.

An example of such a plot is shown in Figure 2 with a quantitative comparison between the levels of a fucosylated triantennary *N*-glycan in 20 ovarian cancer patients (elevated) and age-matched control samples [21]. Statistical comparisons of different glycan profiles suggest that a range of tri- and tetra-antennary glycans, with varying degrees of sialylation and fucosylation, deserve further attention for their distinctly elevated levels in ovarian cancer patients [21]. While this particular comparison appears clinically significant due to its high ROC value, the MS measurements alone do not indicate which of the two possible isomers, if either, is more associated with cancer. However, sample treatment with an appropriate exoglycosidase mixture, followed by MS, can identify a specific isomer [21, 22]. In most other comparisons [21 – 23], the outer-arm fucosylation isomer appears significant.

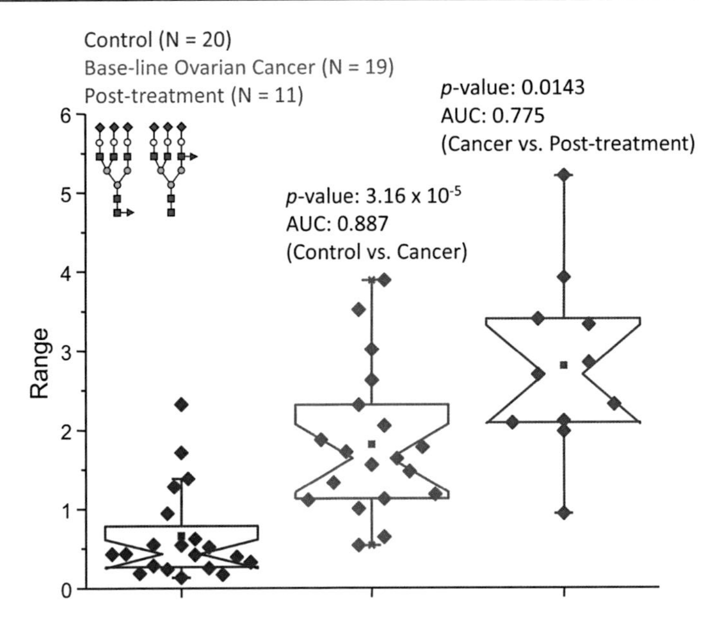

Figure 2 Notched-box plot for a fucosylated triantennary/trisialylated glycan measured through an MS-based profiling; it is not clear from MS, which isomer is involved. After the sample treatment with appropriate exoglycosidase mixture (followed by MS, both isomers were found, but only the glycan with the antennael fucosylation appeared to correlate with the cancer condition). Reprinted with permission from reference [21]. Copyright 2012 American Chemical Society.

However, many of the key *N*-glycans that distinguish healthy from disease samples are mostly evident as trace components whose authenticity must still be confirmed and their exact structures have to be determined. Unfortunately, as yet, these compounds are not synthetically available. Similar glycan types have been observed in the investigations of lung cancer [22] and colorectal cancer [23]. These represent just a few cases of a pressing need for pure glycans to be used as analytical standards.

Analytical Problems of Glycan Isomerism

Whereas a different binding of glycan isomers to a protein is easily envisaged due to different interactions of more or less sterically accessible polar groups, there are limitations to assess sugar isomerism even through techniques as powerful as mass spectrometry. Complementary technologies, such as NMR spectrometry could be additionally informative, but the tiny quantities of sugars extracted from the biological materials of interest limited a reliable structural work. The above-mentioned example of the determination of a fucosyl substitution (Figure 2), whether a core- or branch-substituted, represents only a small sampling of the structural problems to be encountered in the present-day glycobiology.

Yet another biomarker-related glycan isomerism case concerns different linkages of a sialic acid attachment in both simple and multi-antennary glycans. It has been suggested [24] that a conversion between α2,3- and α2,6-linkage could be indicative of certain cancer conditions. The recently developed derivatisation procedure [25] converts α2,3-linked sialic acid residues to lactone structures and α2,6-linked moieties to amides, so that appreciably different molecular masses can be measured through MS. These trends could be seen for a limited number of measurements in breast cancer sera [26], but a bit more complicated scenario is suggested through a comparison of the sets of lung cancer sera and those of control patients with different smoking histories (Figure 3) [22]. While these trends are suggestive of structural complexities concerning these differently sialylated and fucosylated glycans, there is currently no direct way to verify these preliminary findings and assign the position of sialic acids and their substitutions on different antennae for such triantennary glycans. Short of the availability of the authentic synthetic glycans, isolating these compounds from unique biological materials appears desirable for additional structural investigations.

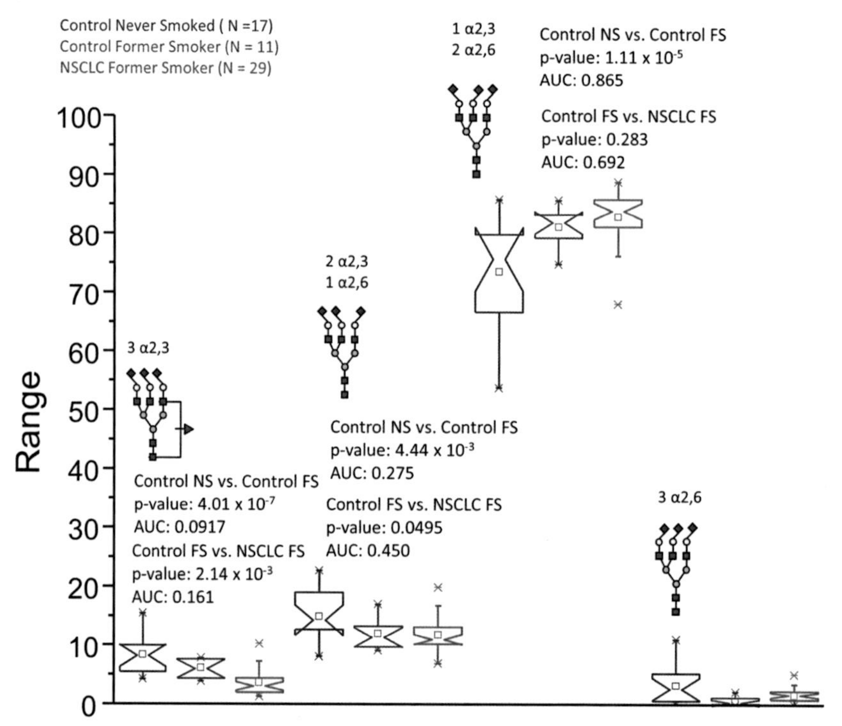

Figure 3. MS measurements of a series of triantennary fucosylated glycans using a derivatisation procedure [25] distinguishing differently sialylated glycans in different patient cohorts. Data adapted from reference [22].
NSCLC = non-small cell lung carcinoma.

Analytical separation science can also make a significant contribution to solve at least some glycan isomerism problems: Some isomers appear separable by special chromatographic columns (through HILIC and graphitised carbon packings) prior to their recording by MS and tandem MS [for a review, see reference 5]. Additionally, capillary electrophoresis (CE) with its exquisite resolving power appears uniquely capable to separate isomeric glycans due to their differences in hydrodynamic radii [27, 28]. To become detectable in CE, carbohydrates must be tagged with a fluorophore prior to their recording by laser-induced fluorescence. Unfortunately, the CE-based separation techniques are not easily combined with MS for a positive identification of glycans at this time. Once again, the availability of authentic electromigration standards could yield substantial progress in structural verification of glycan isomers.

Recycling HPLC

In 1998, Lan and Jorgenson [29] demonstrated the power of recycling HPLC through fully resolving phenylalanine from its deuterated analogue in a 90-min run. Until very recently [30], this methodology has not been applied to the isolation/purification of compounds, including sugars originating from biological sources. To utilise this preparative methodology in glycoscience applications, we decided to investigate several types of UV-absorbing tags as chromophores for detection and performed recycling chromatography through the use of twin columns located on opposite sides of a high-pressure switching valve (Figure 4). This procedure allows the control of optimised "effective column" lengths, so that the analytes experience repeated redirecting of the effluent of one column to the inlet of the other. The overall resolution by the HPLC system is thus controlled in a highly reproducible manner, causing the glycans of closely related structures to become separated from each other and collected at appropriate times. The chromatographic aspects of this procedure are described in a recent publication [30].

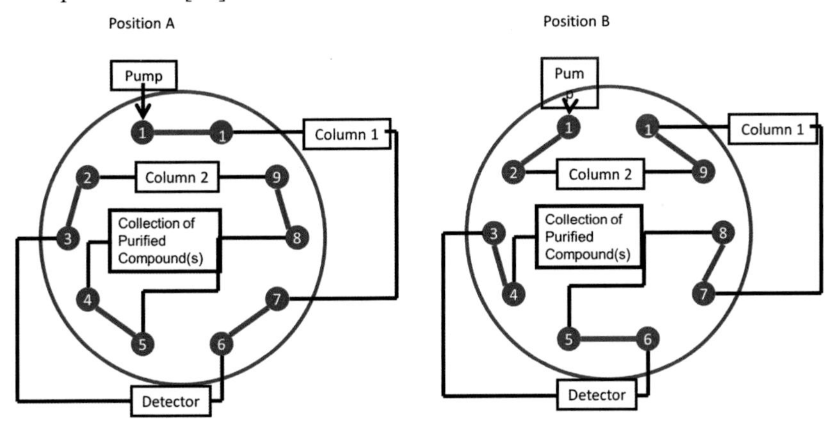

Figure 4. Valve schematics and column configuration used for recycling HPLC employed in glycan purification. Reprinted with permission from reference [30]. Copyright 2013 American Chemical Society.

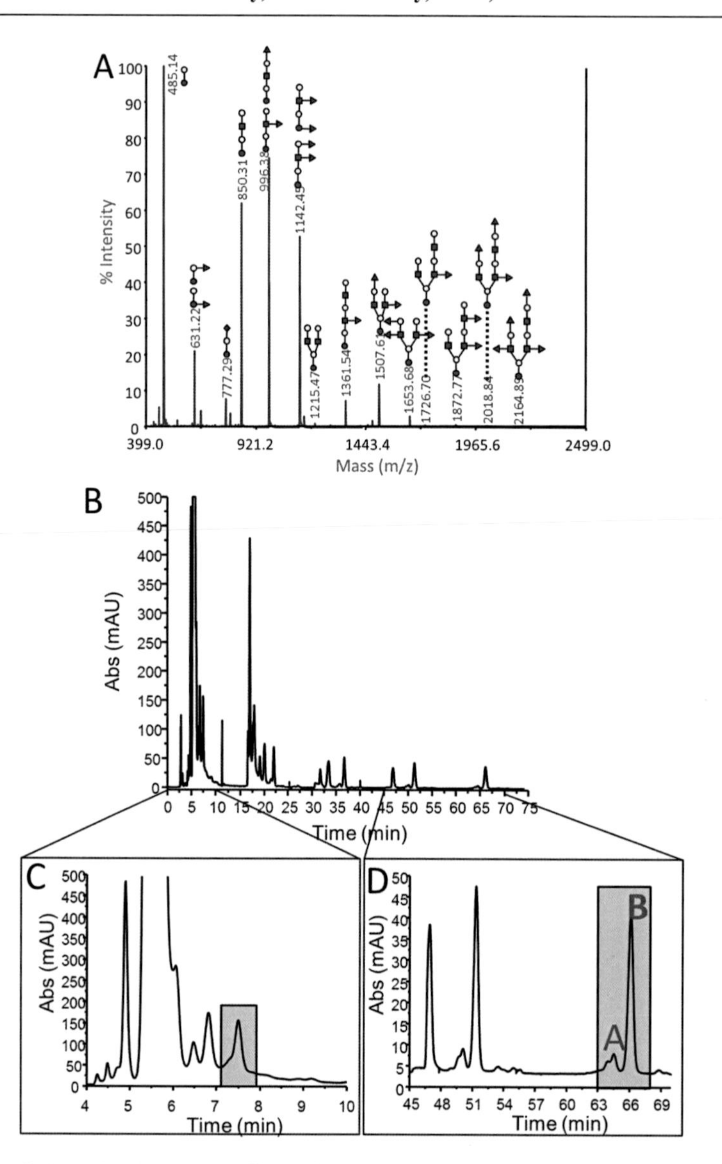

Figure 5. (A) MALDI MS profile of free oligosaccharides present in human breast milk. **(B)** R-HPLC chromatogram for the isolation and purification of human milk oligosaccharides. **(C)** a single chromatographic peak (highlighted) that **(D)** split into two peaks (A and B) due to recycling. Modified and reprinted with permission from reference [30]. Copyright 2013 American Chemical Society.

We have demonstrated the merits of recycling HPLC with the resolution of branching isomers, linear isomers differing in their linkages, and other forms of isomerism [30]. To provide unusual glycan structures to the glycobiology community as analytical standards in microgram-to-milligram amounts, it will be necessary to identify the biological sources with a sufficient abundance of the respective glycoproteins and develop effective bulk isolation strategies and relatively inexpensive glycan cleavage methods prior to the use of the finer separations by HPLC. Among the readily available sources of free oligosaccharides is human breast milk (total carbohydrate content estimated around 5 – 15 mg/mL). Besides its major carbohydrate, lactose, up to 200 different structures with different monosaccharide sequences, linkages, and substitutions have been discovered in this biological material [31]. We have shown [30] how recycling chromatography could be applied to this biological material toward isolation of sugars with closely related structures. Figure 5A shows a brief characterisation of carbohydrates from a pooled breast milk sample (a MALDI-MS profile), while Figure 5B demonstrates a selected cut of the recycling purification in which a further selected fraction (Figure 5C) was further isolated, and subsequently resolved into fractions A and B (Figure 5D). Both peaks, while subsequently characterised by MALDI-MS and MS/MS, indicated the presence of differently fucosylated isomers [30], which could be individually recovered preparatively, and if needed, utilised as pure analytical standards.

An important future use of pure glycans is their attachment to the surfaces of the glycan array assemblies. Although, traditionally, synthetic glycans have been utilised in this capacity, glycans isolated from natural sources can be applied as well. This effort is meant to complement the current activities by others who synthesise various oligosaccharides for the same purpose. Toward the goals of our investigations, we have chosen 4-(2-aminoethyl)-aniline as a molecule with both aromatic and aliphatic amines, with two significantly different pKa values. This allows us to attach the reducing carbohydrate through the reductive amination to the aromatic amine of our chromophore, while the aliphatic amine facilitates attachment to the appropriately derivatised array surface, as seen in Scheme 1.

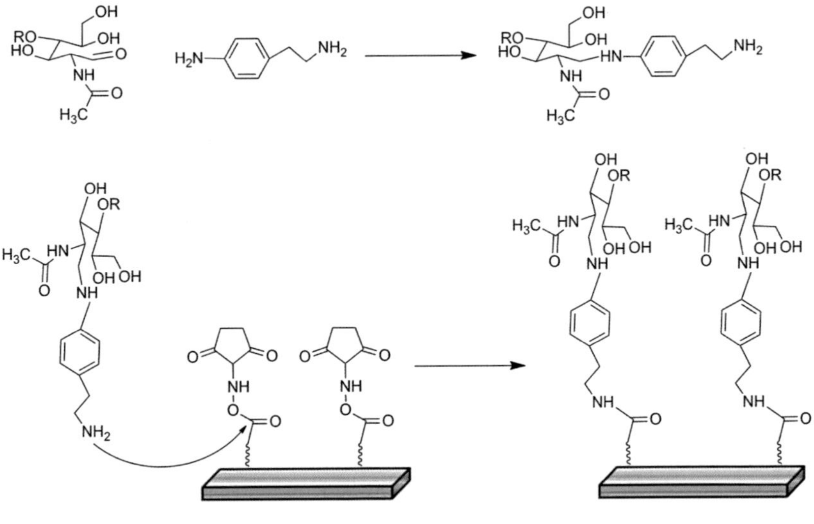

Scheme 1. Tagging procedure to a) detect glycans during R-HPLC and b) attach purified analytes on activated glass surfaces.

A verification of the feasibility of Scheme 1 is demonstrated in Figure 6, where HPLC of tagged glycans derived from human α_1-acid glycoprotein first separates a triantennary/trisialylated glycan and its fucosylated version (Figure 6a), as verified by their mass spectra (Figure 6b and c) and their detection, after immobilisation on an *N*-hydroxysuccinimide-fuctionalised glass slide, by appropriate fluorescently labelled lectins (Figure 6 d and e) [30]. Similarly, we were able to attach high-mannose glycans isolated from ribonuclease B (and resolved through recycling HPLC) in a glycan array-like microdot deposition procedure (Alley, W.R. Jr., Huflejt, M.E., Novotny, M.V., unpublished experiments).

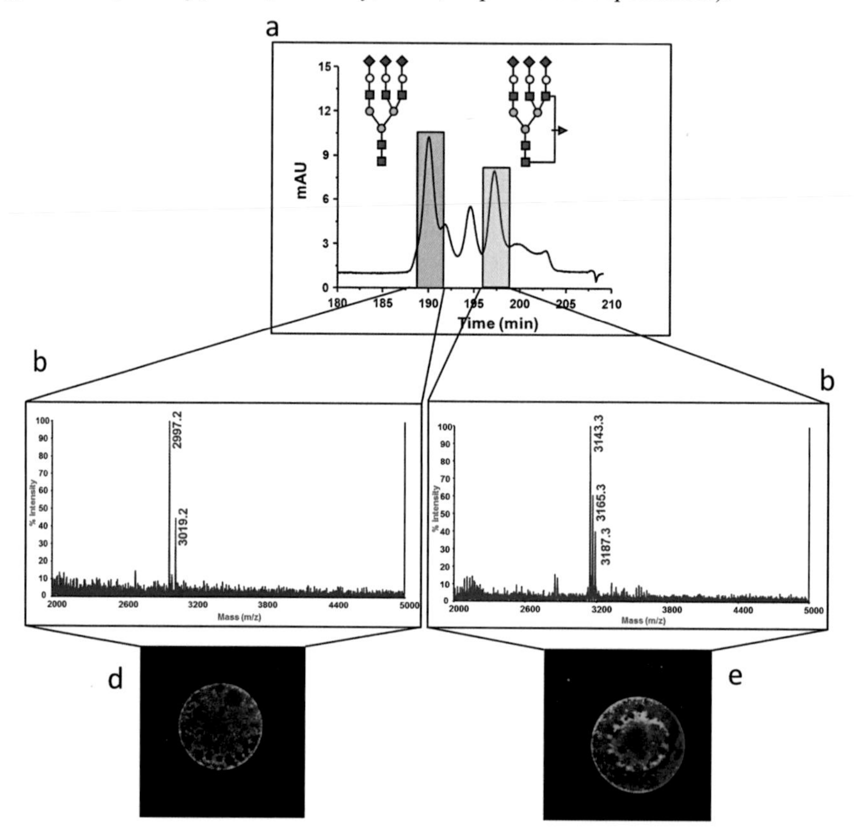

Figure 6. **(a)** R-HPLC chromatogram of a triantennary-trisialylated glycan and its fucosylated analogue derived from human α1-acid glycoprotein; **(b)** and **(c)** negative-mode MALDI mass spectra for the triantennary-trisialylated glycan and its fucosylated version, respectively; **(d)** and **(e)** triantennary-trisialylated glycan and its fucosylated analogue, respectively, immobilised on a glass surface and stained with an appropriate fluorescently-labeled SNA or AAL lectins, respectively. Reprinted with permission from reference [30]. Copyright 2013 American Chemical Society.

Recycling HPLC may also be an attractive purification alternative for synthetically or biosynthetically-derived glycans. While it has been recently demonstrated [15] that very complex structures can be reliably assembled through a sophisticated synthesis, the compounds' purity assessment will likely necessitate a combination of orthogonal analytical techniques. As shown in Figure 7 with a synthesis product, a triantennary glycan with a "mixed population" sialylation (one α2–6 and two α2–3 sialyl substitutions) after 9 column passes, the collected cut is nearly pure, and a single symmetrical peak is seen in the 18th column cut. However, capillary electrophoresis (CE), a technique capable of isomer separations, can still reveal a peak-splitting phenomenon (Mitra, I., Jacobson, S.C., Alley, W.R., Jr., Novotny, M.V., unpublished experiments).

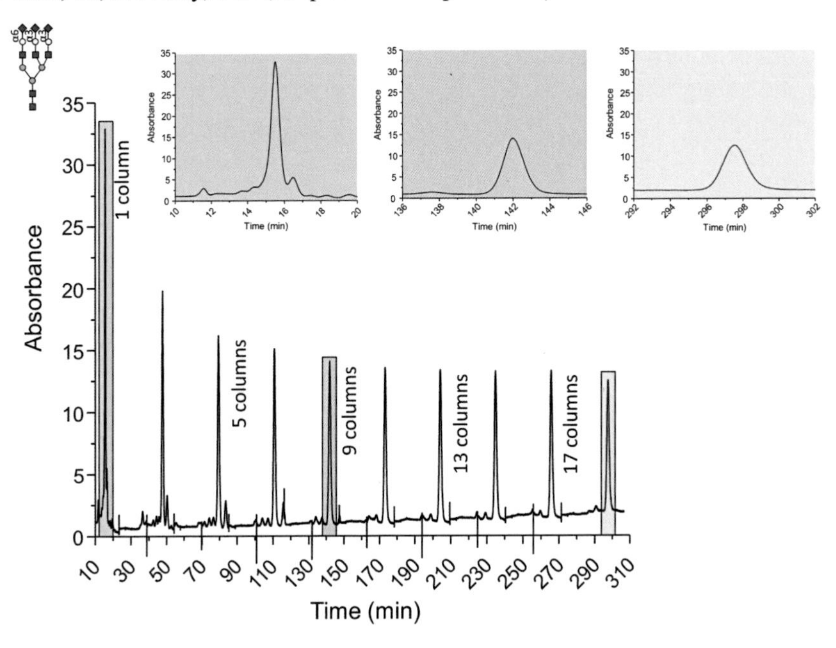

Figure 7. Purification of a synthetic glycan through HILIC recycling HPLC. The sample was obtained through courtesy of Dr. Geert-Jan Boons.

Glycan Isolation from Unusual Materials

Historically, extracting relatively large quantities of glycans from complex biological materials for the sake of structural elucidation was necessary in much of the pioneering studies in glycobiology. Conversely, the last two decades, following the advances in biomolecular MS combined with the continuous column miniaturisation efforts, have led to high-sensitivity glycan profiling capabilities [4], which necessitate only small volumes of physiological fluids. While interesting glycan structures are often suggested by their mass spectra, there is a need for their authentication concerning their isomerism. Once again, isolation from much larger volumes of biological media (containing, unfortunately, numerous "ballast proteins") may become necessary, needing, in turn, larger-scale chromatographic fractionations.

Among the suitable sources of unusual glycan structures, biological exudates (actively secreted fluids of cancer patients) may be strongly considered because (a) they are particularly rich in protein content (> 30 mg/ml concentrations), likely containing the glycoproteins with aberrant glycan structures due to malignancy and inflammation; and (b) large volumes of the fluids are typically treated as waste products in a clinical environment. In one example of a glycan profile (Figure 8) of an ascitic fluid from an ovarian cancer patient, we can observe through MS multiply-fucosylated and sialylated structures. However, these are only tentatively assigned structures without a positive identification of a number of possible isomers. As yet, authentic glycans of this type cannot be produced synthetically, but their isolation and testing for biological activity through a glycan array arrangement becomes a distinct possibility.

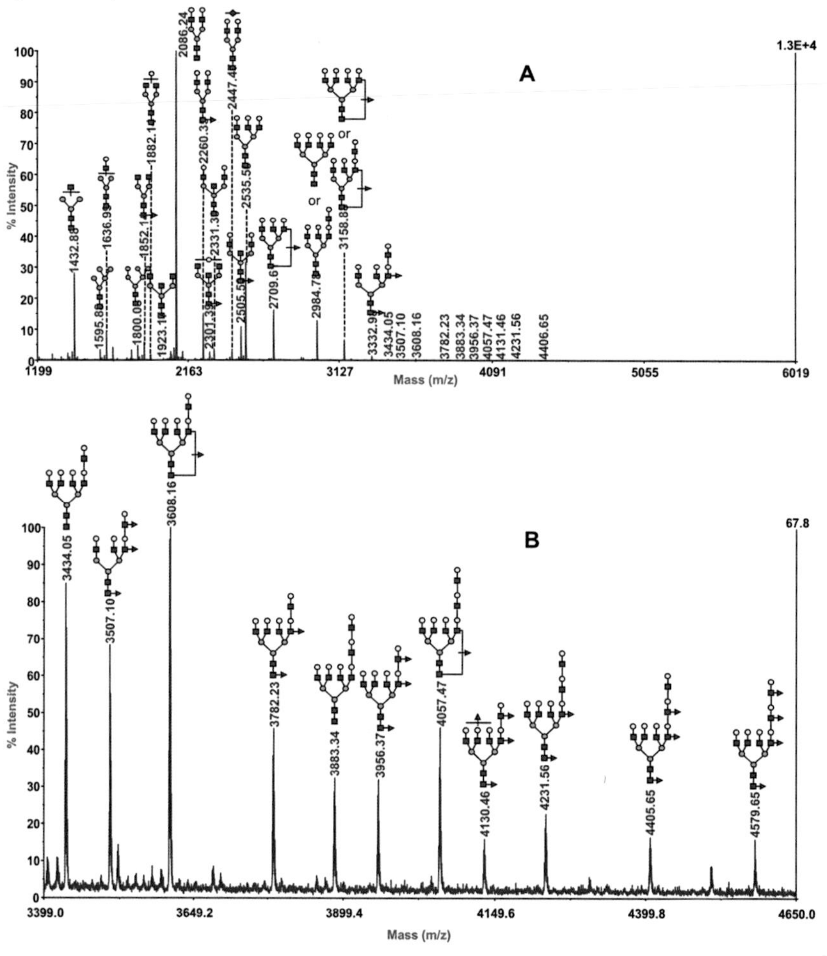

Figure 8. MALDI-MS profile of permethylated N-glycans extracted from glycoproteins of an ascitic fluid sample of an ovarian cancer patient **(A)**; **(B)** represents a high-mass end of the profile. Sample obtained through courtesy of Dr. Daniela Matei, Indiana University, School of Medicine, Indianapolis, IN.

Conclusion

Modern techniques of analytical glycoscience, through its inclusive fields of glycomics, glycoproteomics and glycobioinformatics, have significantly enriched our perception of the ubiquity of glycoconjugate structures in nature and the apparent complexity of their interactions with other biomolecules. While MS has now become a key technology to measure even trace quantities of the easily identifiable glycans, it cannot address directly the isomerism questions of, for example, fucosyl and sialyl substitutions in putative cancer biomarker molecules. A wider availability of authentic glycans for the benefits of glycoscience community must be addressed by efforts in glycan synthesis as well as the isolation and purification of glycans derived from natural sources. The previously under-utilised technique of recycling HPLC has now been demonstrated as a beneficial route to isolate unusual glycans, including isomeric molecules, for the sake of standard availability and future use in glycan array technologies.

Acknowledgements

The authors gratefully acknowledge support of the National Institute of General Medical Sciences under grant R01GM024349 – 27. We wish to thank Dr. Geert-Jan Boons (Complex Carbohydrate Research Center, University of Georgia) for the synthetic glycan samples provided for our studies, Dr. Margaret Huflejt (Department of Cardiothoracic Surgery, Langone Medical Center, New York University) for valuable discussions concerning the needs of glycan array fabrication, and Dr. Daniela Matei (Indiana University, School of Medicine, Simon Cancer Center, Indianapolis) for providing the samples of ovarian cancer ascitic fluids.

References

[1] Laine, R.A. (1997) Information capacity of the carbohydrate code. *Pure and Applied Chemistry* **69**:1867 – 1873.
doi: 10.1351/pac199769091867.

[2] Gabius, H.J. (2009) The sugar code: fundamentals of glycosciences. Wiley-VCH, John Wiley: Weinheim, Chichester.
http://www.wiley.com/WileyCDA/WileyTitle/productCD-352732089X.html.

[3] Cummings, R.D. (2009) The repertoire of glycan determinants in the human glycome. *Molecular BioSystems* **5**:1087 – 10104.
doi: 10.1039/b907931a.

[4] Novotny, M.V., Alley, W.R., Jr., Mann, B.F. (2013) Analytical glycobiology at high sensitivity: current approaches and directions. *Glycoconjugate Journal* **30**:89 – 117.
doi: 10.1007/s10719-012-9444-8

[5] Alley, W.R., Jr., Mann, B.F., Novotny, M.V. (2013) High-sensitivity analytical approaches for the structural characterization of glycoproteins. *Chemical Reviews* **113**:2668 – 2732.
doi: 10.1021/cr3003714

[6] Novotny M.V., Alley, W.R., Jr. (2013) Recent trends in analytical and structural glycobiology. *Current Opinion in Chemical Biology* **17**: 832 – 841.
doi: 10.1016/j.cbpa.2013.05.029

[7] Transforming Glycoscience: A Roadmap for the Future (2012) National Research Council of the National Academies, The National Academy Press, Washington, D.C. http://www.nap.edu/catalog.php?record_id=13446.

[8] Rillahan, C.D., Paulson, J.C. (2011) Glycan Microarrays for Decoding the Glycome. *Annual Review of Biochemistry* **80**:797 – 823.
doi: 101146/annurev-biochem-061809-152236.

[9] Liu, Y. Palma, A.S., Feizi, T. (2009) Carbohydrate microarrays: key developments in glycobiology. *Biological Chemistry* **390**: 647 – 656.
doi: 101515/BC.2009.071.

[10] Smith, D.F., Song, X., Cummings, R.D. (2010) Use of glycan microarrays to explore specificity of glycan-binding proteins. *Methods in Enzymology* **480**:417 – 444.
doi: 10.1016/S0076-6879(10)80033-3.

[11] Vuskovic M.I., Xu, H., Bovin, N.V., Pass, H.I., Huflejt, M.E. (2011) Processing and analysis of serum antibody binding signals from Printed Glycan Arrays for diagnostic and prognostic applications. *International Journal of Bioinformatics Research and Applications* **7**:402 – 426.
doi: 10.1504/IJBRA.2011.043771.

[12] Park, S., Gildersleeve, J.C., Blixt, O., Shin, I. (2013) Carbohydrate microarrays. *Chemical Society Reviews* **44**:4310 – 4326.
doi: 10.1039/c2cs35401b.

[13] Lepenies, B., Yin, J., Seeberger, P.H. (2010) Applications of synthetic carbohydrates to chemical biology. *Current Opinion in Chemical Biology* **14**:404 – 411.
doi: 10.1016/j.cbpa.2010.02.016.

[14] Schmalz, R.M., Hansson, S.R., Wong, C.-H. (2011) Enzymes in the synthesis of glycoconjugates. *Chemical Reviews* **111**:4259 – 4307.
doi: 10.1021/cr200113w.

[15] Wang, Z., Chinoy, Z.S., Ambre, S.G., Peng, W., McBride, R., de Vries, R.P., Glushka, J., Paulson, J.C., Boons, G.-J. (2013) A general strategy for the chemo-enzymatic synthesis of asymmetrically branched *N*-glycans. *Science* **341**:379 – 383. doi: 10.1126/science.1236231.

[16] Varki, A., Freeze, H.H., Glycans in Acquired Human Diseases. In *Essentials of Glycobiology* (2009 2nd ed.; Varki, A., Cummings, R.D., Esko, J.D., Freeze, H.H., Stanley, P., Bertozzi, C.R., Hart, G.W., Etzler, M.E., Eds.: Cold Spring Harbor (NY). http://www.ncbi.nlm.nih.gov/books/NBK1946/

[17] Taniguchi, N. (2008) Human disease glycomics/proteome initiative (HGPI). *Molecular & Cellular Proteomics* **7**:626 – 627. http://www.mcponline.org/content/7/3/626.full?sid=09972842-78ba-4578-8d6f-8949eb867f7c

[18] Grivennikov, S.I., Greten, F.R., Karin, M. (2010) Immunity, Inflammation, and Cancer. *Cell* **140**:883 – 899. doi: 10.1016/j.cell.2010.01.025.

[19] Narimatsu, H., Sawaki, H., Kuno, A., Kaji, H., Ito, H., Ikehara, Y. (2009) A strategy for discovery of cancer glycol-biomarkers in serum using newly developed technologies for glycoproteomics. *The FEBS Journal* **277**:95 – 105. doi: 10.1111/j.1742-4658.2009.07430.x.

[20] Kyselova, Z., Mechref, Y., Al Bataineh, M., Dobrolecki, L.E., Hickey, R.J., Sweeney, C.J., Novotny, M.V. (2007) Alterations in the serum glycome due to metastatic prostate cancer. *Journal of Proteome Research* **6**:1822 – 1832. doi: 10.1021/pr060664t.

[21] Alley, W.R. Jr, Vasseur, J.A., Goetz, J.A., Svoboda, M., Mann, B.F., Matei, D.E., Menning, N., Hussein, A., Mechref, Y., Novotny, M.V. (2012) *N*-linked glycan structures and their expressions change in the blood sera of ovarian cancer patients. *Journal of Proteome Research* **11**: 2282 – 2300. doi: 10.1021/pr201070k.

[22] Vasseur, J.A., Goetz, J.A., Alley, W.R., Jr., Novotny, M.V.(2012) Smoking and lung cancer-induced changes in *N*-glycosylation of blood serum proteins. *Glycobiology* **22**:1684 – 1708. doi: 10.1093/glycob/cws108.

[23] Alley, W.R., Jr., Svoboda, M., Goetz, J.A., Vasseur, J.A. Novotny, M.V. (2014) Glycomic analysis of sera derived from from colorectal cancer patients reveals increased fucosylation of highly-branched *N*-glycans, submitted to publication.

[24] Orntoft, T.F., Vestergaard, E.M. (1999) Clinical aspects of altered glycosylation of glycoproteins in cancer. *Electrophoresis* **20**:362 – 371.
doi: 10.1002/(SICI)1522-2683(19990201)20:2<362::AID-ELPS362>3.0.CO;2-V.

[25] Alley, W.R., Jr., Novotny, M.V. (2010) Glycomic analysis of sialic acid linkages in glycans derived from blood serum glycoproteins. *Journal of Proteome Research* **9**: 3062 – 3072.
doi: 10.1021/pr901210r.

[26] Alley, W.R., Jr., Madera, M., Mechref, Y., Novotny, M.V. (2010) Chip-based reversed-phase liquid chromatography-mass spectrometry of permethylated *N*-linked glycans: a potential methodology for cancer-biomarker discovery. *Analytical Chemistry* **82**:5095 – 5106.
doi: 10.1021/ac10013.

[27] Zhuang, Z., Starkey, J.A., Mechref, Y., Novotny, M.V., Jacobson, S.C. (2007) Electrophoretic analysis of *N*-glycans on microfluidic devices. *Analytical Chemistry* **79**:7170 – 7175.
doi: 10.1021/ac071261v.

[28] Mitra, I., Alley, W.R. Jr., Goetz, J.A., Vasseur, J.A., Novotny, M.V., Jacobson, S.C. (2013) Comparative profiling of *N*-glycans isolated from serum samples of ovarian cancer patients and analyzed by microchip electrophoresis. *Journal of Proteome Research* **12**:4490 – 4496.
doi: 10.1021/pr400549e.

[29] Lan, K., Jorgenson, J.W. (1998) Pressure-induced retention variations in reversed-phase alternate-pumping recycle chromatography. *Analytical Chemistry* **70**: 2773 – 2782.
doi: 10.1021/ac971226w.

[30] Alley, W.R., Jr., Mann, B.F., Hruska, V., Novotny, M.V. (2013) Isolation and purification of glycoconjugates from complex biological sources by recycling high-performance liquid chromatography. *Analytical Chemistry* **85**:1048 – 10416.
doi: 10.1021/ac4023814.

[31] Zivkovic, A.M., German, J.B., Lebrilla, C.B., Mills, D.A. (2011) Human milk glycobiome and its impact on the infant gastrointestinal microbiota. *Proceedings of the National Academy of Sciences of the United States of America* **108** Suppl 1: 4653 – 4658.
doi: 10.1073/pnas.1000083107.

Discovering the Subtleties of Sugars
June 10th – 14th, 2013, Potsdam, Germany

Towards Identifying Protective Carbohydrate Epitopes in the Development of a Glycoconjugate Vaccine against *Cryptococcus neoformans*

Lorenzo Guazzelli and Stefan Oscarson*

Centre for Synthesis and Chemical Biology, University College Dublin, Belfield, Dublin 4, Ireland

E-Mail: *stefan.oscarson@ucd.ie

Received: 10th February 2015 / Published: 16th February 2015

Cryptococcus neoformans

Cryptococcus neoformans is an opportunistic encapsulated yeast that causes cryptococcal meningoencephalitis (cryptococcosis) in immunocompromised individuals, including AIDS patients [1], organ transplant recipients [2] or other patients receiving immunosuppressive drugs. Infection with *C. neoformans* is acquired by inhalation of desiccated yeast cells into the lungs, which causes a local pulmonary infection. The yeast cells can enter the bloodstream and disseminate to the skin, bone and the central nervous system, thereby causing a systemic infection. The pathogen is able to cross the blood-brain-barrier, the mechanism of which is not fully understood yet [3]. Once inside the brain the pathogen destroys the surrounding tissue [1]. Studies showed that most adults in New York City have antibodies against *C. neoformans* [4] but cryptococcosis is a relatively rare disease in immunocompetent individuals despite the widespread occurrence of *C. neoformans* in the environment. Presumably, immunocompetent individuals are able to mount an immune response without showing any clinical symptoms of a cryptococcal infection. Epidemiological studies indicate that *C. neoformans* remains dormant in the host, and that cryptococcosis may be the result of re-activation of a latent infection [5]. It was suggested that cryptococcal infection occurs in childhood [6, 7], and that childhood infection may predispose people to airway diseases, such as asthma, later in life [8]. During the past four decades the number of immunocompromised people increased due to the AIDS epidemic, which in turn led to a dramatic rise in fungal infections [1].

This article is part of the Proceedings of the Beilstein Glyco-Bioinformatics Symposium 2013.
www.proceedings.beilstein-symposia.org

It is estimated that *C. neoformans* infects at least 1 million AIDS patients worldwide annually, which results in 650,000 deaths each year [9]. Currently, cryptococcosis is treated with antifungal drugs [10]. However, the use of antifungals is problematic because they are generally highly toxic, and have only a limited ability to eradicate an infection. Furthermore, excessive use of fungal drugs facilitates the emergence of resistant strains [11], why there is a strong need for a vaccine against *C. neoformans*.

The cell wall of *C. neoformans* is surrounded by a polysaccharide capsule, which is a main virulence factor and thus a potential target for the development of a capsular polysaccharide based vaccine. The cryptococcal capsule consists of two major polysaccharides, glucuronoxylomannan (GXM) and galactoxylomannan (GalXM). GXM comprises 90 – 95% of the total capsule mass and GalXM approximately 5 – 8%. In addition, a minor amount of mannoprotein (< 1%) is present in the capsule [12].

Bacterial Glycoconjugate Vaccines

Glycoconjugate vaccines based on functionalized bacterial capsular polysaccharides conjugated to a carrier protein has proven to be excellent and safe vaccines and glycoconjugate vaccines against bacteria causing meningitis (*Haemophilus influenzae*, *Neisseria meningitidis*, and *Streptococcus pneumoniae*) are now commercial and introduced into mass vaccination schemes in many countries [13 – 15]. Many of these bacterial polysaccharides are also immunogenic in their free unconjugated form and have been used as vaccines since the 1970 s but the conjugation to a protein in the glycoconjugate vaccines leads to a T-cell dependent immune response with many added advantages including prolonged immunity through formation of memory cells, possibility to boost the immune response by repeated injections, and, perhaps most important, activity also in small children (under 18 months of age).

Carbohydrate structures are common on the surface of microbe cells, mainly procaryotic bacteria cells, but also parasite and fungi cells. Bacterial surface polysaccharides are of two types, either a capsular (exo) polysaccharide (CPS) or a lipopolysaccharide (LPS, in Gram-negative bacteria), the latter containing fatty acid residues anchoring the structure in the outer cell membrane. Both structures show a large structural variety but also strain and group specificity [16]. These structures are of main importance for the virulence of the bacteria since it protects against dehydration and phagocytosis. Generally, the bacterial polysaccharides are built up by a repeating unit (1 – 10 monosaccharides) which is polymerized through glycosidic or phosphodiester linkages. Hence the structure is homogeneous along the chain and between different batches of the polysaccharide and an NMR spectrum of the polysaccharide looks (more or less) like a spectrum of the repeating unit, which facilitates structural analysis.

Glycoconjugate vaccines based on synthetic oligosaccharides have been an area of research for a longer time [15, 17] and recently a commercial vaccine based on synthetic oligosaccharides corresponding to *Haemophilus influenzae* type b was developed and licensed [18]. This vaccine was found to be as efficient as the ones based on the bacterial polysaccharide and is now used in mass vaccination schemes. Due to the structural complexity of many bacterial polysaccharide a glycoconjugate vaccine containing synthetic oligosaccharides is most often not commercially cost-effective, but, since through synthesis any part oligosaccharide structure of the polysaccharide is available, they make excellent research tools for structure-activity relationship studies of the immune response, for example, to investigate the (smallest) size and structure of oligosaccharide that will give a protective immune response (a protective epitope) [19]. Most often these studies have been carried out in mice and it is important to recognize that there are major differences between the mice and the human immune system, why the size and structure of protective epitopes in humans are probably quite different from the ones established in mice.

Cryptococcus neoformans GXM CPS structure

The structure of fungal polysaccharides are quite different from bacterial polysaccharides, they are not built up from repeating units, but are heteropolymers where only structural motifs can be elucidated and the ratio between them established (much like plant polysaccharides) but no definite structure given. Furthermore, different batches of polysaccharides contain different ratios between the structural motifs why reproducibility is a major issue when considering the use of them in a vaccine. The heterogeneity has its origin in the capsule biosynthesis, but this is much less studied in fungi than the corresponding biosynthesis of bacterial CPS [20]. The current structural model of GMX (which was established already in the 1980 s) involves six structural motifs ('triads') based on mannose trimers (Figure 1) [12, 21]. *Cryptococcus neoformans* is divided into four serotypes, A, B, C, and, D, and it has been possible to (at least partly) correlate these to the structure of the suggested "triads" [22]. Strains of serotype A and D are the most frequent cause of cryptococcosis in humans and thus the serotypes of primary interest for a human vaccine [1]. The basic structural motif consists of an α-D-(1→3)-mannopyranan backbone which is substituted with a β-D-glucopyranosyluronic acid residue at OH-2 of the first mannosyl residue of the triad. The mannan backbone can be further substituted with β-D-xylopyranosyl residues at OH-2, and/or with β-D-xylopyranosyl residues at OH-4, and the different amount of xylose substitution defines the different serotypes.

In addition, some of the hydroxyl groups of the GXM polysaccharide are esterified with acetyl groups adding more heterogeneity to the structures [23, 24]. The degree of *O*-acetylation varies from serotype to serotype; serotypes A and D have the highest, whereas B and C have the lowest degree of acetylation. An average of two acetyl groups per triad is found for serotypes A and D. ^{13}C NMR analysis of a polysaccharide of serotype A revealed that the acetyl groups are most likely located at OH-6 of the backbone mannosyl residues [23].

A lack of virulence in mutants where the acetyl transfer enzyme had been knocked-out indicates that the acetyl group is essential for virulence and hence probably also immunologically relevant [25].

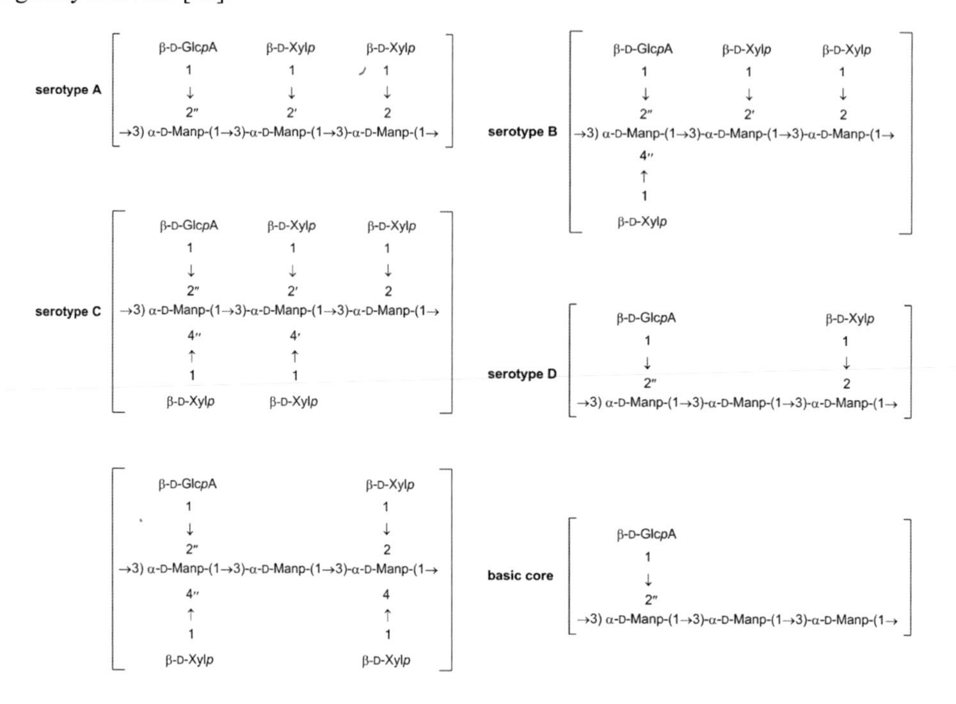

Figure 1. Suggested structural motifs of GXM.

Candidate Vaccines Based on Native Capsular Polysaccharides

Phagocytic cells are important in host defense against microbial pathogens. They ingest foreign material that has been opsonized by antibodies and/or complement [26]. The polysaccharide capsule of *C. neoformans* has anti-phagocytic properties, and as a consequence the cryptococcal cells are able to evade killing by phagocytes [27]. The rationale behind vaccination against *C. neoformans* is to elicit antibodies that can opsonize the fungal cells, and thereby facilitate their clearance through subsequent phagocytosis [28, 29]. In 1958, Gadebusch performed the first immunization studies using whole killed *Cryptococci* cells [30, 31]. However, these vaccines were unsuccessful in protecting mice against experimental cryptococcosis. Vaccines that used attenuated live *Cryptococci* cells as immunogens gave encouraging results [31]. It was observed that immunised mice survived significantly longer than non-immunised mice after inoculation with *Cryptococci* cells but use of whole cell vaccines is not optimal why continued efforts focused on part structure vaccines. Most of these studies involved GXM, the major constituent of the capsule. In the 1960 s, Goren and Middlebrook produced the first glycoconjugate vaccine, which was composed of unfractionated GXM polysaccharide conjugated to bovine gamma globulin [32]. The vaccine was highly immunogenic, but did not give a protective antibody immune response. In 1991, Devi *et al.* developed another glycoconjugate vaccine, which consisted of fractionated GXM polysaccharide conjugated to tetanus toxoid (TT) [33]. The GMX-TT conjugate vaccine was again highly immunogenic and both active and passive immunization of mice conferred protection against experimental cryptococcosis [34, 35]. However, further studies by Casadevall *et al.* showed (by investigating a library of created monoclonal antibodies) that the GXM-TT vaccine did not only elicit protective (neutralize the fungi), but also non-protective (bind but do not kill the fungi) and even deleterious (disease-enhancing) antibodies [36, 37]. Moreover, it was shown that the free un-conjugated GXM polysaccharide, in contrast to many bacterial polysaccharides, had potent immunosuppressive properties [38 – 40], which further complicated its use as a vaccine component.

These results, which more or less disqualify the use of native GXM polysaccharide in vaccine development, were interpreted to be a consequence of the micro-heterogeneity of the GXM polysaccharide and represent a major difference when compared to bacterial CPS-based vaccines [41]. A hypothesis to explain the immunological results was proposed suggesting that there are protective epitopes within the GXM polysaccharide which when part of a glycoconjugate vaccine will produce a protective antibody response, but also, due to the heterogeneity, that there are non-protective epitopes within the GXM polysaccharide which when part of a glycoconjugate vaccine will produce a non-protective antibody response, which might also prevent the action of formed protective antibodies. The major question is how to identify the protective as well as the non-protective epitopes present in the heterogeneous native polysaccharide? In spite of having access to a library of both protective and non-protective mAbs, there was no knowledge at all about their binding specificities. A crystal structure of one of the protective antibodies has been published, but only with a peptide, mimicking the native carbohydrate substrate, in the antigen binding site [62]. Considering the heterogeneity (and mostly unknown biosynthesis) of the native CPS, arguably the only way to identify the different types of epitopes as well as to produce protective epitopes to be used in vaccine development is through chemical synthesis of well-defined part structures of the GXM (again differing from bacterial CPSs where usually the native polysaccharide is a possible (and often better) alternative).

Synthetic Approaches to GXM Capsular Polysaccharide Structures

From a synthetic perspective, the preparation of fragments of the GXM capsular polysaccharide including the (probably immunogenically important) acetyl groups in the targets is a challenge [42]. Esters are often used in carbohydrate chemistry not only as temporary protecting groups of hydroxyl functions, but also as participating groups during glycosylation reactions. Installing an acetate or a benzoate in the 2-position of the donor ensures high selectivity in the formation of 1,2-transglycosidic linkages. Considering that all the sugars of the GXM CPS are linked by 1,2-transglycosidic linkages, it is evident that excluding esters from the set of possible protecting groups is a substantial limitation. A further complication are difficulties with low yields experienced in benzylation of glucuronic acid residues, which complicates the strategy to use acyl protecting groups during the glycosylation step (to ensure 1,2-trans selectivity) followed by change of acyl protecting groups to benzyl groups and a final introduction of the 6-*O*-acetyl group. This works well for the xylose-containing building blocks but with glucuronic acid containing blocks the benzylation reaction is low-yielding and with reproducibility problems especially on a larger scale. Thus, alternative pathways are required.

Figure 2. Desired building blocks.

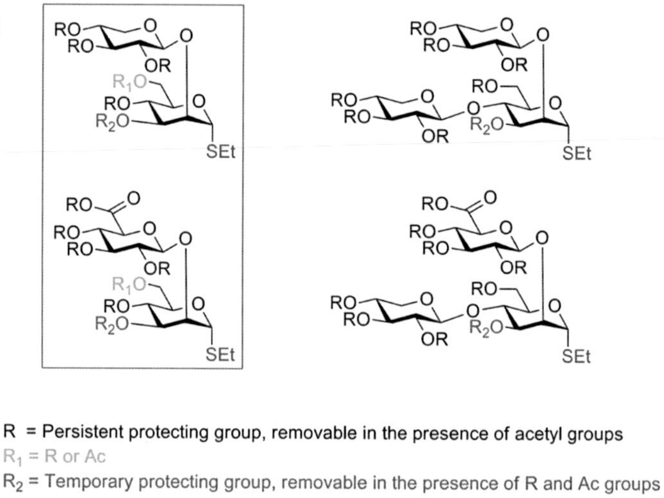
R = Persistent protecting group, removable in the presence of acetyl groups
R1 = R or Ac
R2 = Temporary protecting group, removable in the presence of R and Ac groups

Scheme 1. Failed building block glycosylation attempt.

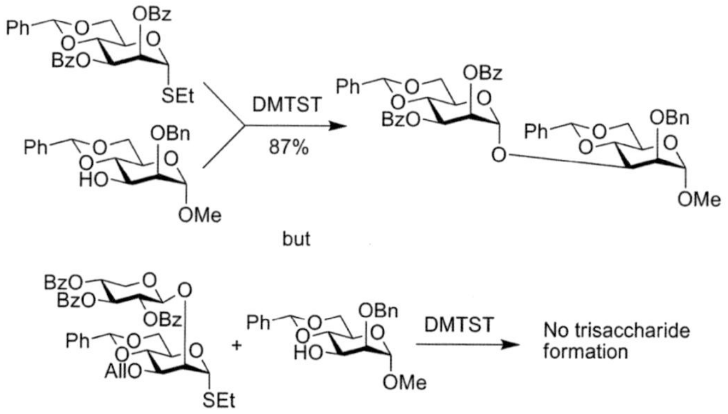
DMTST
87%
but
DMTST
No trisaccharide formation

In the structural motifs suggested for the GXM polysaccharide (Figure 1) two disaccharides, β-D-GlcA-(1→2)-α-D-Man and β-D-Xyl-(1→2)-α-D-Man, and two trisaccharides, β-D-GlcA-(1→2)-[β-D-Xyl-(1→4)]-α-D-Man and β-D-Xyl-(1→2)-[β-D-Xyl-(1→4)]-α-D-Man, can be identified as common part structures, why a convergent synthetic strategy based on these as building blocks would probably be the most efficient way to produce GXM oligosaccharides (Figure 2). Initially, we had major problems in applying this strategy, since when using a thiodisaccharide β-D-Xyl-(1→2)-α-D-Man block as donor in couplings to a mannose acceptor no trisaccharides were obtained (Scheme 1).

Because of these problems a linear approach was instead investigated (Scheme 2). This work was performed in collaboration with Robert Cherniak at Georgia State University and the oligosaccharides were designed to be used as inhibitors of the binding of native GXM to antibodies, why they were synthesized as their methyl glycosides [43 – 45]. Acetyl groups were introduced as the last step into some already deprotected oligosaccharides why there were no issue with using acyl protecting groups (to facilitate 1,2-*trans* stereoselectivity in the glycosylations).

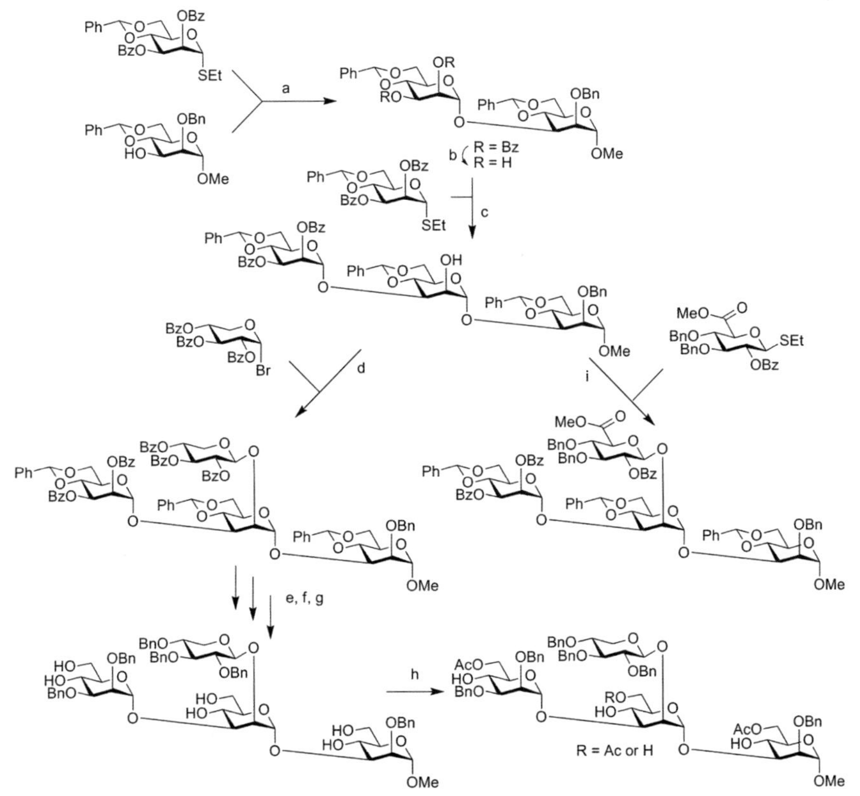

Scheme 2. Synthesis of tetrasaccharide fragments of GXM serotype A using a linear strategy. Reagents and conditions: **(a)** DMTST, DCM, MS 4 Å, 20 °C, 2 h, 87%; **(b)** 1. NaOMe, DCM/MeOH (1:1), 20 °C, o/n; 2. Dowex® H^+ ion-exchange resin; **(c)** 1) Bu_2SnO, MeOH, reflux, 30 min; 2) DMTST, DCM, MS 4 Å, 20 °C, 2 h 20 min, 40% over three steps; **(d)** AgOTf, DTBP, DCM, MS 4 Å, - 35 °C, 2 h, 79%; **(e)** 1. NaOMe, DCM/MeOH (1:2), 20 °C, o/n; 2. Dowex® H^+ ion-exchange resin; **(f)** NaH, BnBr, DMF, 0 °C→20 °C, 2 h, 75% over two steps; **(g)** AcOH, MeCN, 65 °C, 4 h; **(h)** acetyl chloride, sym-collidine, DCM, - 70 °C→20 °C, 3 h 20 min, 60% 10a, 15% 10b; **(i)** DMTST, DCM, MS 4 Å, 20 °C, o/n, 30%.

A small number of di- to tetrasaccharides was synthesized but none showed any activity in following inhibition experiments.

At about the same time van Boom *et al.* published a synthesis of a pentasaccharide corresponding to GXM serotype D, as its methyl glycoside and without acetates, employing a convergent approach. In this synthesis a glucose donor was used to avoid problems often encountered with glucuronic acid donors, and the glucuronic acid moiety was formed by oxidation of OH-6 after the glycosylation sequence (Scheme 3) [46].

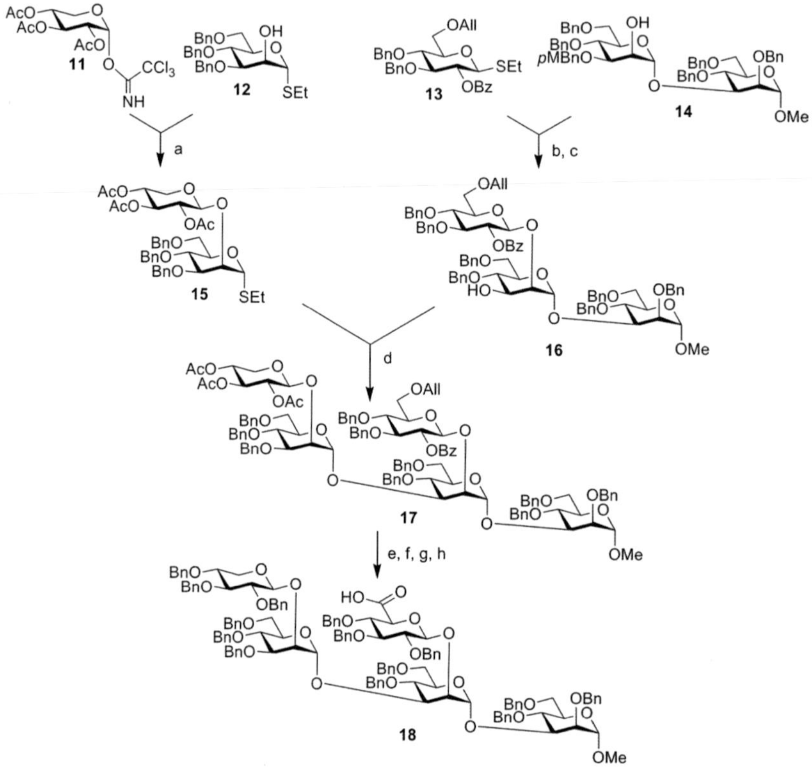

Scheme 3. Synthesis of a pentasaccharide fragment of GXM serotype D via convergent strategy and 'post-glycosylation oxidation' approach. Reagents and conditions: **(a)** TMSOTf, 1,2-dichloroethane, -40 °C→-20 °C, 5 h, 38%; **(b)** NIS, TfOH, 1,2-dichloroethane-Et_2O, MS 4 Å, -30 °C, 15 min, 81%; **(c)** DDQ, dichloromethane-water, 1 h, 82%; **(d)** NIS, TfOH, Et_2O, MS 4 Å, 0 °C, 30 min, 67%; **(e)** KOtBu, MeOH, 20 h, 73%; **(f)** BnBr, NaH, DMF, 3 h, 88%; **(g)** 1. Ir(COD)$[PCH_3(Ph)_2]_2PF_6$, 1,2-dichloroethane, 70 h; 2. HCl-MeOH (0.5 M), 22 h, 78%; **(h)** 1. oxalyl chloride, DCM, DMSO, - 60 °C, 90 min; 2. $NaClO_2$, 2-methyl-2-butene, NaH_2PO_4, *t*-BuOH, H_2O, 20 h, 17 h.

Similar non-acetylated methyl glycoside structures of serotype A [47 – 49], serotype B [50, 51], and serotype C [52] were prepared by Kong *et al.* following a mixed convergent-linear strategy. Synthesis of a heptasaccharide structural motif of GXM serotype B is described in Scheme 4, as an example of this approach. To our knowledge there are no reports in the literature of the use of any of these synthetic structures (either van Boom's or Kong's) in following biological experiments.

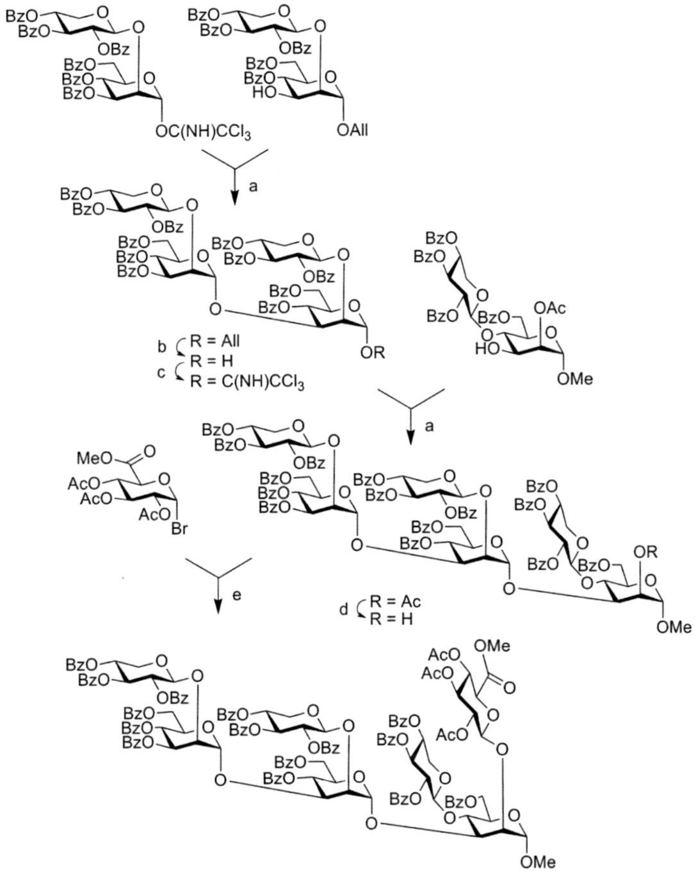

Scheme 4. Synthesis of a heptasaccharide fragment of GXM serotype B via mixed convergent-linear strategy. Reagents and conditions: **(a)** TMSOTf, CH_2Cl_2, -20 °C→ 0 °C, 90% for 21, 70% for 26; **(b)** $PdCl_2$, MeOH, 4 h, 89%; **(c)** CCl_3CN, DBU, CH_2Cl_2, 3 h, 89%; **(d)** MeCOCl/MeOH/CH_2Cl_2, 48 h, 45%; **(e)** 2,4-lutidine, AgOTf, CH_2Cl_2, -20 °C→ 0 °C, 4 h, 78%.

In a collaboration with Prof. Casadevall at the Albert Einstein College of Medicine, New York, involving structures to be parts of a glycoconjugate vaccine candidate, larger structures containing (the believed important) 6-*O*-acetyl groups of the mannan backbone as well as a spacer to allow conjugation to a carrier protein was targeted and the block synthetic approach revisited. Since mainly serotype A and D (no 4-*O*-Xyl substituent) were of interest disaccharide building blocks were synthesized. It was found that the earlier glycosylation problems encountered (Scheme 1) could easily be avoided by opening of the 4,6-*O*-benzylidene acetal in the used donor, which also gave the possibility to introduce the desired 6-*O*-acetyl group. Still the issue of introducing a benzyl protected glucuronic acid moiety remained, but different pathways were investigated and optimized, a "post-glycosylation oxidation approach" (Scheme 5) and one using a benzylated glucuronic acid donor, to allow efficient synthesis of gram quantities of this disaccharide [53, 54].

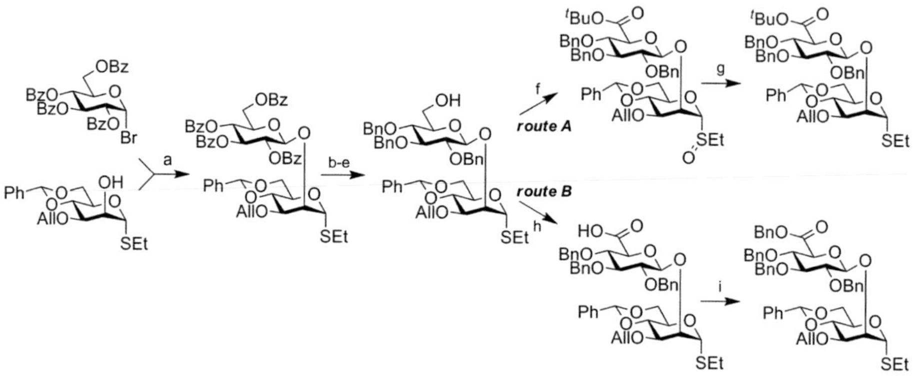

Scheme 5. Synthesis of a glucuronic acid-containing disaccharide using the 'post-glycosylation oxidation' approach. Reagents and conditions: **(a)** AgOTf, DTBP, DCM, MS 4 Å, - 40 °C, o/n, 92%; **(b)** 1. NaOMe, MeOH, 20 °C, o/n; 2. Dowex® H^+ ion-exchange resin, 69%; **(c)** DMTrCl, pyridine, 20 °C, 3 h; **(d)** NaH, BnBr, DMF, 0 ° C→20 °C, 1 h; **(e)** *p*-TsOH, DCM/MeOH (2:1), 0 °C, 20 min, 43% (over three steps); **(f)** PDC, Ac_2O, *t*-BuOH, DCM, 20 °C, o/n; **(g)** Ph_3P, I_2, NaI, MeCN, 20 °C, 2 h (30% over three steps); **(h)** TEMPO, BAIB, DCM/H_2O (2:1), 20 °C, 2 h, 81%; **(i)** $PhCHN_2$, EtOAc, 20 °C, 2 h, 76%.

With these building blocks in hand their couplings to give larger structures proved to be unproblematic, DMTST as promoter in diethyl ether gave high yields of glycosylation products with complete stereo-selectivity even without the use of 2-*O*-participating group (which was expected considering that α-mannoside linkages were being formed) and structures up to the size of a heptasaccharide were efficiently synthesized (Scheme 6) [55 – 57]. By the use of either 6-*O*-acetylated or non-acetylated building blocks in some of the glycosylations, three different acetylation patterns were introduced in the final deprotected heptasaccharide, one with two acetates (6'- and 6'''-), one with one acetate (6'''-), and one with no acetates [57].

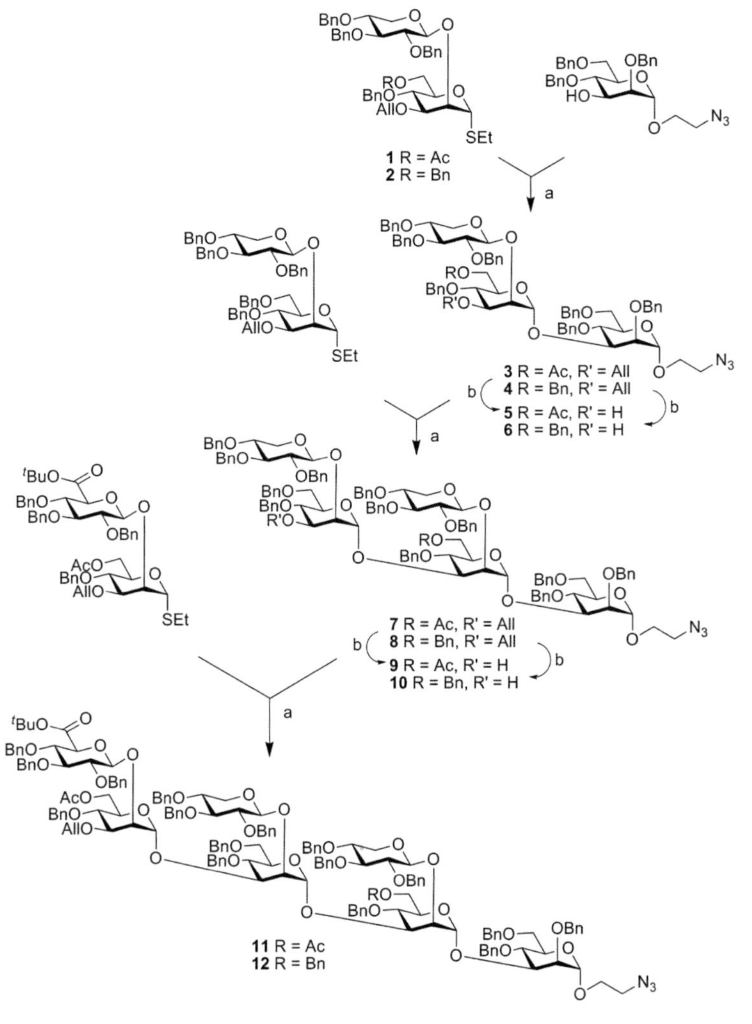

Scheme 6. Synthesis of heptasaccharide fragments of GXM serotype A via convergent strategy and 'pre-glycosylation oxidation' approach. Reagents and conditions: **(a)** DMTST, Et_2O, 65% for 3, 91% for 4, 65% for 5, 85% for 6, 70% for 7, 70% for 8; **(b)** $PdCl_2$, MeOH-EtOH, 60% for 9, 73% for 10; 80% for 11, 62% for 12.

Glycoconjugate Vaccine Candidates Based on Synthetic GXM Oligosaccharides

The amino-containing spacer part of this heptasaccharide structure now allowed both direct ELISA binding studies with monoclonal antibodies as well as formation of a protein conjugate for immunological evaluation. The structures (with different acetylation pattern) were conjugated to biotin and the conjugates obtained fixated onto Streptavidin-coated ELISA plates and screened against Prof. Casadevall's library of mAbs [57]. The 6'''-*O*-mono-acetylated heptasaccharide was also conjugated to human serum albumin (HSA) using the squarate ester methodology and used in mice immunization studies (Scheme 7) [57, 58].

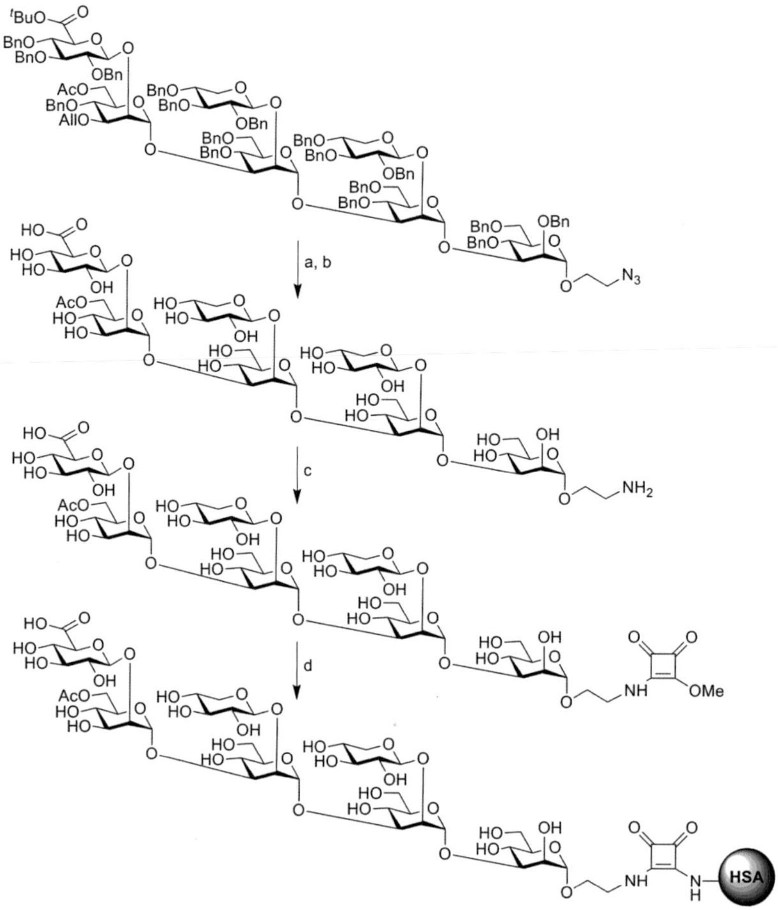

Scheme 7. Synthesis of a synthetic candidate vaccine. Reagents and conditions: **(a)** $PdCl_2$, EtOH/MeOH, 2 h, 74%; **(b)** $Pd(OH)_2$, H_2 (8 atm), EtOAc:AcOH:H_2O (4:1:1), 48 h, 93%; **(c)** Dimethyl squarate, MeOH, Et_3N, 4 h; **(d)** HAS, Labasco buffer, 24 h.

For the first time biological activity was found with our synthetic GXM oligosaccharides. In the ELISA tests seven mAbs of the library recognized the heptasaccharide, more or less with the same affinity for all three acetylation pattern investigated (Figure 3). In the immunization study, using Freund's complete adjuvant, high antibody titers were obtained in the immunized mice (4 with 10 µg/dose, 4 with 2.5 µg/dose, Figure 4) as compared to controls and the antibodies were confirmed to be mainly of the IgG isotype (IgG1, IgG2a, IgG2b) and to bind to native GXM. However, when the immunized mice were challenged with the fungi they died. Also, the mAbs that recognized the heptasaccharide were all IgM and non-protective antibodies.

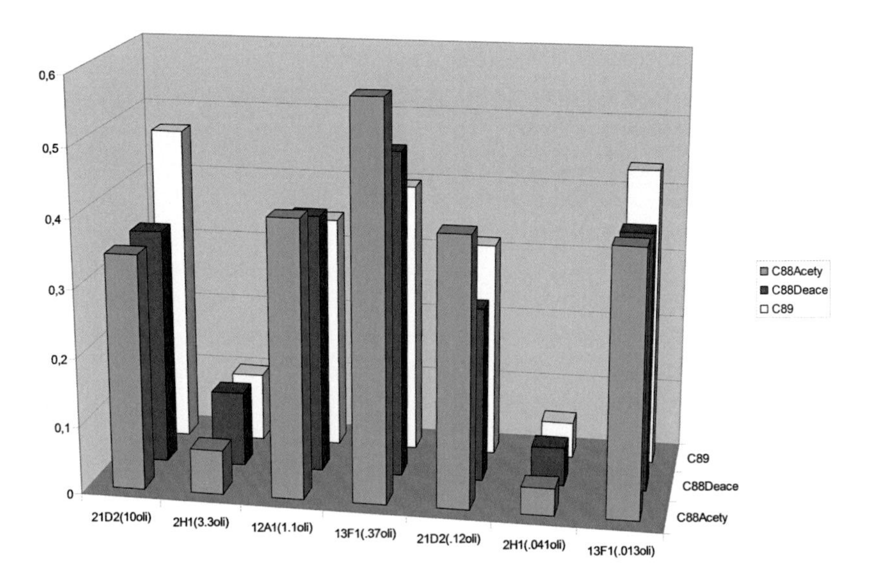

Figure 3. mAb Binding study (C88 6',6'''-di-OAc; C89 6'''-mono-OAc).

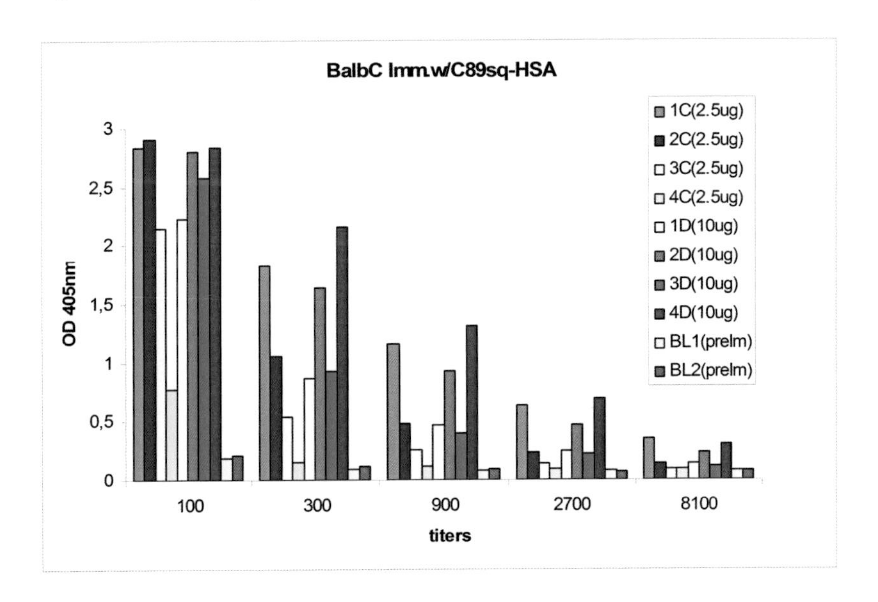

Figure 4. Antibody titres in immunization study.

Although a bit disappointing these results show that the working hypothesis is correct, and proves one part of it, there are non-protective GXM oligosaccharide structures that when used in a vaccine give rise to a non-protective immune response.

In the (continuing) quest to identify protective epitopes it is quite difficult if not impossible to really predict anything about size and structure. Bundle and co-workers working on the CPS of the fungi *Candida albicans* have identified a trisaccharide as a promising optimal epitope to include in a vaccine and also found larger structures to be less effective [59, 60]. But in the *Cryptococcus* case so far no structures of sizes up to a heptasaccharide have been found to be protective. There are speculations that protective epitopes in native GMX polysaccharide may be conformational, thus that the carbohydrate component has to be of sufficient size/length in order to form immunologically relevant secondary structures [57]. To address this possibility, we are now both looking into modelling the polysaccharide to investigate if (and if so when and which) secondary structures are formed as well as synthesizing an extended library of GXM structures containing both more structures and longer structures.

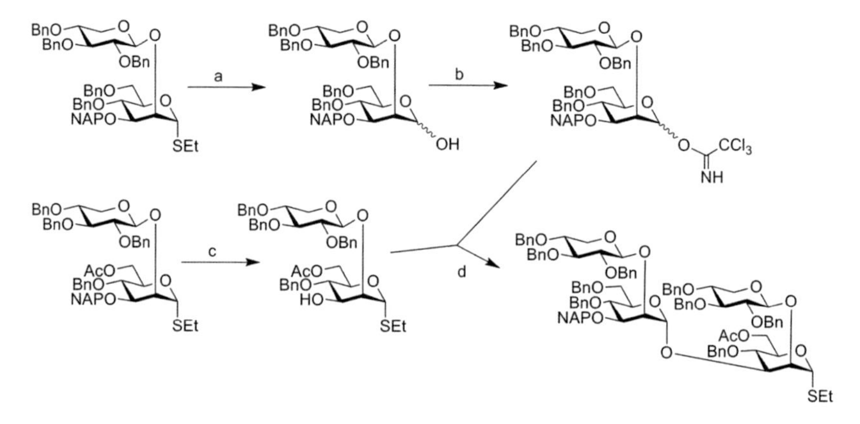

Scheme 8. Synthesis of a Type A tetrasaccharide building block. Reagents and conditions: **(a)** NIS, TFA, DCM, H_2O, 85%; **(b)** DBU, Cl_3CCN, DCM, quant.; **(c)** DDQ, DCM, H_2O, 75%; **(d)** TBDMSOTf, toluene, -30 °C, 95%.

For longer GXM structures, construction of larger building blocks than the disaccharides so far used would greatly facilitate their synthesis, why continued synthesis of tri- and penta-saccharide building blocks for serotype D and tetra- and hexasaccharide building blocks for serotype A have been pursued. By changing slightly the protecting group pattern of the original disaccharide building blocks and optimizing glycosylation conditions we have been able to produce both the tri- and tetrasaccharide blocks discussed (Scheme 8) [61] as well as the penta- and hexasaccharide building blocks and use them in the construction of a library of spacer-containing GXM structures ranging from mono- to octadecasaccharides and with variant acetylation pattern, which is now ready for screening with the library of mAbs and also to be conjugated to a carrier protein and used in mice immunizations.

Summary

Due to the heterogeneity and often low immunogenicity of fungal capsular polysaccharides their possible use in the development of a glycoconjugate vaccine is much complicated as compared to the use of bacterial polysaccharides for the same purpose. Results from mice immunization experiments using protein conjugates of the native *Cryptococcus neoformans* GXM polysaccharide showed variable results with formation of both protective, non-protective and deleterious antibodies, demonstrating the problems encountered with these heterogeneous fungal polysaccharides. To investigate which GXM structures that give rise to the different types of antibodies and to identify protective epitopes to be used in a Cryptococcus glycoconjugate vaccine an approach using well-defined synthetic part structures of the GXM polysaccharide has been instigated. With the help of these synthetic structures a non-protective epitope has been identified proving the viability of the approach. With an extended library of synthetic GXM oligosaccharide structures the search is now continuing to identify also protective epitopes.

Acknowledgements

We thank Science Foundation Ireland (Grants 08/IN.1/B2067 and 13/IA/1959) and Marie Curie-Intra-European Fellowships (FP7-PEOPLE-2011-IEF, project number 299710) for financial support.

References

[1] Mitchell, T.G., Perfect, J.R. (1995) *Clin. Microbiol. Rev.* **8**:515 – 548.

[2] Husain, S., Wagener, M.M., Singh, N. (2001) *Emerg. Infect. Dis.* **7**:375 – 381. doi: 10.3201/eid0703.017302.

[3] Del Poeta, M., Casadevall, A. (2012) *Mycopathologia* **173**:303 – 310. doi: 10.1007/s11046-011-9473-z.

[4] Currie, B.P., Casadevall, A. (1994) *Clin. Infect. Dis.* **19**:1029 – 1033. doi: 10.1093/clinids/19.6.1029.

[5] Garcia-Hermoso, D., Janbon, G., Dromer, F.J. (1999) *Clin. Microbiol.* **37**: 3204 – 3209.

[6] Goldman, D.L., Khine, H., Abadi, J., Lindenberg, D.J., Pirofski, L.-A., Niang, R., Casadevall, A. (2001) *Pediatrics* **107**:E66. doi: 10.1542/peds.107.5.e66.

[7] Abadi, J., Pirofski, L.-A. (1999) *J. Infect. Dis.***180**:915 – 919. doi: 10.1086/314953.

[8] Arora, S., Huffnagle, G. (2005) *Immunol. Res.* **33**:53 – 68.
doi: 10.1385/IR:33:1:053.

[9] Park, B.J., Wannemuehler, K.A., Marston, B.J., Govender, N., Pappas, P.G., Chiller, T.M. (2009) *Aids* **23**:525 – 530.
doi: 10.1097/QAD.0b013e328322ffac.

[10] Brizendine, K.D., Pappas, P.G. (2010) *Curr. Infect. Dis. Rep.* **12**:299 – 305.
doi: 10.1007/s11908-010-0113-4.

[11] Barker, K., Rogers, P.D. (2006) *Curr. Infect. Dis. Rep.* **8**:449 – 456.
doi: 10.1007/s11908-006-0019-3.

[12] Cherniak, R., Sundstrom, J.B. (1994) *Infect. Immun.* **62**:1507 – 1512.

[13] Ravenscroft, N., Jones C. (2000) *Curr. Opin. Drug Discovery Dev.* 222 – 231.

[14] Ada, G., Isaacs, D. (2003) *Clin. Microbiol. Infect.* **9**:79 – 85.
doi: 10.1046/j.1469-0691.2003.00530.x.

[15] Costantino, P., Rappuoli, R., Berti F. (2011) *Expert Opin. Drug Discov.* **6**: 1045 – 1066.
doi: 10.1517/17460441.2011.609554.

[16] Lindberg, B., Kenne, L. (1985) In G.O. Aspinall (Ed.), *The Polysaccharides*, Vol. 2, Academic Press, New York, pp. 287 – 363.

[17] Astronomo, R.D., Burton, D.R. (2010) *Nature Reviews Drug Discovery* **9**(4): 308 – 324.
doi: 10.1038/nrd3012.

[18] Verez-Bencomo, V. *et al.* (2004) *Science* **305**:522 – 525.
doi: 10.1126/science.1095209.

[19] See *e.g.* Benaissa-Trouw, B., Lefeber, D.J., Kamerling, J.P., Vliegenthart, J.F.G., Kraaijeveld, K., Snippe, H. (2001) *Infect. Immun.* **69**:4698 – 4701 (doi: 10.1128/IAI.69.7.4698-4701.2001) and Safari, D. *et al.* (2008) *Infect. Immun.* **76**:4615 – 4623. (doi: 10.1128/IAI.00472-08).

[20] Doering, T.L. (2000) *Trends Microbiol.* **8**:547 – 553.
doi: 10.1016/S0966-842X(00)01890-4.

[21] Cherniak, R., Jones, R.G., Reiss, E. (1988) *Carbohydr. Res.* **172**:113 – 138.
doi: 10.1016/S0008-6215(00)90846-2.

[22] Cherniak, R., Reiss, E., Slodki, M.E., Plattner, R.D., Blumer, S.O. (1980) *Mol. Immunol.* **17**:1025 – 1032.
doi: 10.1016/0161-5890(80)90096-6.

[23] Bhattacharjee, A.K., Bennett, J.E., Glaudemans, C.P.J. (1984) *Rev. Infect. Dis.* **6**:619 – 624.
doi: 10.1093/clinids/6.5.619.

[24] Cherniak, R., Valafar, H., Morris, L.C., Valafar, F. (1998) *Clin. Diagn. Lab. Immunol.* **5**:146 – 159.

[25] Ellerbroek, P.M., Lefeber, D.J., van Veghel, R., Scharringa, J., Brouwer, E., Gerwig, G.J., Janbon, G., Hoepelman, A.I.M., Coenjaerts, F.E.J. (2004) *J. Immunol.* **173**:7513 – 7520.
doi: 10.4049/jimmunol.173.12.7513.

[26] Kindt, T.J., Goldsby, R.A., Osborne, B.A. (2007) *Kuby Immunology*; 6th ed. W.H. Freeman and Company: New York.

[27] García-Rodas, R., Zaragoza, O. (2012) *FEMS Immunol. Med. Microbiol.* **64**: 147 – 161.
doi: 10.1111/j.1574-695X.2011.00871.x.

[28] Cutler, J.E., Deepe Jr, G.S., Klein, B.S. (2007) *Nat. Rev. Microbiol.* **5**:13 – 28.
doi: 10.1038/nrmicro1537.

[29] Casadevall, A., Pirofski, L.-A. (2005) *Med. Mycol.* **43**:667 – 680.
doi: 10.1080/13693780500448230.

[30] Gadebusch, H.H. (1958) *J. Infect. Dis.* **102**:219 – 226.
doi: 10.1093/infdis/102.3.219.

[31] Segal, E. (1987) *Crit. Rev. Microbiol.* **14**:229 – 273.
doi: 10.3109/10408418709104440.

[32] Goren, M.B., Middlebrook, G.M. (1967) *J. Immunol.* **98**:901 – 913.

[33] Devi, S.J., Schneerson, R., Egan, W., Ulrich, T.J., Bryla, D., Robbins, J.B., Bennett, J.E. (1991) *Infect. Immun.* **59**:3700 – 3007.

[34] Devi, S.J.N. (1996) *Vaccine* **14**:841 – 844.
doi: 10.1016/0264-410X(95)00256-Z.

[35] Casadevall, A., Mukherjee, J., Devi, S.J.N., Schneerson, R., Robbins, J.B., Scharff, M.D. (1992) *J. Infect. Dis.* **165**:1086 – 1093.
doi: 10.1093/infdis/165.6.1086.

[36] Mukherjee, J., Scharff, M.D., Casadevall, A. (1992) *Infect. Immun.* **60**:4534 – 4541.

[37] Mukherjee, J., Nussbaum, G., Scharff, M.D., Casadevall, A. (1995) *J. Exp. Med.* **181**:405 – 409.
doi: 10.1084/jem.181.1.405.

[38] Vecchiarelli, A. (2000) *Med. Mycol.* **38**:407 – 417.
doi: 10.1080/mmy.38.6.407.417.

[39] Yauch, L.E., Lam, J.S., Levitz, S.M. (2006) *PLoS Path.* **2**:1060 – 1068.
doi: 10.1371/journal.ppat.0020120.

[40] Vecchiarelli, A. (2007) *Crit. Rev. Immunol.* **27**:547 – 557.
doi: 10.1615/CritRevImmunol.v27.i6.50.

[41] Zaragoza, O., Rodrigues, M.L., De Jesus, M., Frases, S., Dadachova, E., Casadevall, A., Allen I. Laskin, S.S., Geoffrey, M.G. (2009) In *Adv. Appl. Microbiol.*; 1st ed. Laskin, A.I., Sima, S., Gadd, G.M., Eds. Academic Press: San Diego, Burlington, London, Vol. 68, p 133.

[42] Morelli, L., Poletti, L., Lay, L. (2011) *Eur. J. Org. Chem.* **29**:5723 – 5777.
doi: 10.1002/ejoc.201100296.

[43] Garegg, P.J., Olsson, L., Oscarson, S. (1993) *J. Carbohydr. Chem.* **12**:955 – 967.
doi: 10.1080/07328309308020108.

[44] Garegg, P.J., Olsson, L., Oscarson, S. (1996) *Biorg. Med. Chem.* **4**:1867 – 1871.
doi: 10.1016/S0968-0896(96)00168-X.

[45] Garegg, P.J., Olsson, L., Oscarson, S. (1997) *J. Carbohydr. Chem.* **16**:973 – 981.
doi: 10.1080/07328309708005731.

[46] Zegelaar-Jaarsveld, K., Smits, S.A.W., van der Marel, G.A., van Boom, J.H. (1996) *Biorg. Med. Chem.* **4**:1819 – 1832.
doi: 10.1016/S0968-0896(96)00164-2.

[47] Zhang, J., Kong, F. (2003) *Carbohyr. Res.* **338**:1719 – 1725.
doi: 10.1016/S0008-6215(03)00264-7.

[48] Zhang, J., Kong, F. (2003) *Tetrahedron Lett.* **44**:1839 – 1842.
doi: 10.1016/S0040-4039(03)00119-9.

[49] Zhang, J., Kong, F. (2003) *Bioorg. & Med. Chem.* **11**:4027 – 4037.
doi: 10.1016/S0968-0896(03)00391-2.

[50] Zhao, W., Kong, F. (2004) *Carb. Res.* **339**:1779 – 1786.
doi: 10.1016/j.carres.2004.04.010.

[51] Zhao, W., Kong, F. (2005) *Bioorg. & Med. Chem.* **13**:121 – 130.
doi: 10.1016/j.bmc.2004.09.049.

[52] Zhao, W., Kong, F. (2005) *Carbohydr. Res.* **340**:1673 – 1681.
doi: 10.1016/j.carres.2005.05.003.

[53] Vesely, J., Rydner, L., Oscarson, S. (2008) *Carbohydr. Res.* **343**:2200 – 2208. doi: 10.1016/j.carres.2007.11.026.

[54] Guazzelli, L., Ulc, R., Oscarson, S. (2014) *Carbohydr. Res.* **389**:57 – 65. doi: 10.1016/j.carres.2014.01.022.

[55] Alpe, M., Oscarson, S., Svahnberg, P. (2003) *J. Carbohydr. Chem.* **22**:565 – 577. doi: 10.1081/CAR-120026459.

[56] Alpe, M., Oscarson, S., Svahnberg, P. (2004) *J. Carbohydr. Chem.* **23**:403 – 416. doi: 10.1081/CAR-200040114.

[57] Oscarson, S., Alpe, M., Svahnberg, P., Nakouzi, A., Casadevall, A. (2005) *Vaccine* **23**:3961 – 3972. doi: 10.1016/j.vaccine.2005.02.029.

[58] Nakouzi, A., Zhang, T., Oscarson, S., Casadevall, A. (2009) *Vaccine* **27**:3513 – 3518. doi: 10.1016/j.vaccine.2009.03.089.

[59] Lipinski, T., Wu, X., Sadowska, J., Kreiter, E., Yasui, Y., Cheriaparambil, S., Rennie, R., Bundle, D.R. (2012) *Vaccine* **30**:6263 – 6269. doi: 10.1016/j.vaccine.2012.08.010.

[60] Johnson, M.A., Bundle, D.R. (2013) Chem. Soc. Reviews **42**:4327 – 4344. doi: 10.1039/c2cs35382b.

[61] Guazzelli, L., Ulc, R., Rydner, L., Oscarson, S., manuscript submitted.

[62] Young, A.C.M., Valadon, P., Casadevall, A., Schaff, M.D., Sacchettini, J.C. (1997) *J. Mol. Biol.* **274**:622 – 634. doi: 10.1006/jmbi.1997.1407.

Biographies

Kiyoko F. Aoki-Kinoshita

received her B.S. and M.S. degrees in computer science from Northwestern University simultaneously in 1996. She received her doctorate in computer engineering from Northwestern in 1999. From 2000, she was employed at BioDiscovery, Inc., in Los Angeles as a senior software engineer before moving to Kyoto, Japan, in 2003 to work as a postdoctoral researcher at the Bioinformatics Center, Institute of Chemical Research, Kyoto University. There she developed various algorithmic and data-mining methods for analyzing glycan structure data accumulated in the KEGG GLYCAN database, which have been published in numerous journal papers. In 2006, she joined the Department of Bioinformatics in the Faculty of Engineering at Soka University in Tokyo and is now an associate professor of bioinformatics. She is also involved in several research projects pertaining to the prediction of glycan functions based on their structure as well as recognition patterns of glycan structures by other proteins and even viruses. One of these projects is the development of a Web resource called RINGS (Resource for INformatics of Glycomes at Soka), which provides many of the informatics algorithms and methods for the glycosciences that have been published in the literature. These tools are all made available freely to allow glycoscientists to take advantage of bioinformatics tools pertinent for their data.

Sabine L. Flitsch

graduated with a Diplom in Chemistry from the University of Münster (Germany) in 1982. She then obtained a Michael Wills Scholarship to study at the University of Oxford under the supervision of Sir Jack Baldwin and obtained her DPhil in 1985. She spent three years at the Massachusetts Institute of Technology, USA as a Research Fellow (with H. G. Khorana) before returning to the UK to hold academic positions at the Universities of Exeter, Oxford, Edinburgh and Manchester. She is currently Professor of Biological Chemistry at the University of Manchester with her research group housed at the Manchester Institute of Biotechnology (MIB).

Her research interests are on the interface of chemistry and biology with focus on using biocatalysis for the synthesis of complex carbohydrates and glycoconjugates such as glycolipids, glycoproteins, polysaccharides and glycomaterials. Her group has developed a broad toolset in these areas ranging from chemical synthesis and physical analytical techniques to protein chemistry and biochemistry with the aim to develop effective biocatalysts for applications in glycoscience.

This article is part of the Proceedings of the Beilstein Glyco-Bioinformatics Symposium 2013.
www.proceedings.beilstein-symposia.org

Sabine currently chairs the Euroglycosciences Forum (EGSF.org) which aims to promote glycosciences within Europe.

Martin G. Hicks

is a member of the board of management of the Beilstein-Institut. He received an honours degree in chemistry from Keele University in 1979. There, he also obtained his PhD in 1983 studying synthetic and theoretical approaches to the photochemistry of pyridotropones under the supervision of Gurnos Jones. He then went to the University of Wuppertal as a post-doctoral fellow, where he carried out research with Walter Thiel on semi-empirical quantum chemical methods. In 1985, Martin joined the computer department of the Beilstein-Institut where he worked on the Beilstein Database project. His subsequent activities involved the development of cheminformatics tools and products in the areas of substructure searching and reaction databases.

Thereafter, he took on various roles for the Beilstein-Institut, including managing directorships of subsidiary companies and was head of the funding department 2000 – 2007. He joined the board of management in 2002; his current interests and responsibilities range from organization of Beilstein Symposia with the aim of furthering interdisciplinary communication between chemistry and neighbouring scientific areas, to the publishing of the Beilstein Open Access journals – *Beilstein Journal of Organic Chemistry* and *Beilstein Journal of Nanotechnology* – and production of scientific videos for Beilstein TV.

Peter Hufnagel

studied chemistry in Hamburg and Munich. In 1997 he received his Doktor degree working on macroscopic Bacteriorhodopsin preparations in the group of Dieter Oesterhelt at the Max Planck Institute of Biochemistry in Martinsried, Germany.

After being a postdoctoral fellow at the Hokkaido National Industrial Research Institute in Sapporo, Japan in 1998, he joined Bruker Daltonics in Bremen, Germany. He worked on the automation of MALDI mass spectra acquisition workflows, and on the development of automats for picking 2D gel spots, in-gel digestion and MALDI sample preparation. He is currently heading Bruker Daltonics' bioinformatics software development team. His main interests are the identification and quantitation of proteins including PTMs, and the development of integrated, user-friendly software.

Niclas Karlsson

The research on post translational modification has been focused on the dynamics of the intracellular phosphorylation. The role of phosphorylation as regulator of cellular events is undisputed, but there is an obvious need for transmitting intracellular events to extracellular mediators Dr Karlsson's group is exploring the role of sulfated oligosaccharides in extracellular signaling in inflammation/infection using model systems and clinical samples and utilizing sophisticated high sensitive mass spectrometry. Mucin type sulfation is most famously known in the homing of lymphocytes to peripheral lymph nodes mediated by sulfated oligosaccharides on special endothelium developed during inflammation called high endothelial venules.

The initial binding of the lymphocytes to sulfated oligosaccharides is via l-selectin. However, other examples for the biological role of sulfation on mucosal surface glycoconjugates are scarce, but this is more a reflection of the lack of analytical tools then lack of biological importance. Locking into rheumatoid arthritis and osteoarthritis we found a sulfomucin identified as Lubricin in synovial fluid and we are currently exploring its role in these diseases.

In order to explore the role of glycosylation for extracellular signalling we are using mass spectrometry and are developing mass spectrometric tools and databases. The development of UniCarb-DB is a collaboration of researchers from the University of Gothenburg, Sweden; Macquarie University (Biomolecular Frontiers Research Centre), Australia; Proteome Informatics Group (Swiss Institute for Bioinformatics) and the National Institute for Bioprocessing Research and Training (NIBRT), Ireland.

Carsten Kettner

studied biology at the University of Bonn and obtained his diploma at the University of Göttingen. In 1999, he was awarded his Ph.D for his work on the biophysical comprehension of the yeast vacuolar ATPase using the patch-clamp techniques in the group of Adam Bertl at the University of Karlsruhe. As a post-doctoral student he continued both the studies on the biophysical properties of the pump and the investigation of the kinetics and regulation of the plasma membrane potassium channel (TOK1). In 2000 he moved to the Beilstein-Institut.

Here, he is responsible (a) for the organization of the Beilstein symposia and the publication of the proceedings of the symposia and (b) for the administration and project management of funded research projects such as the Beilstein Endowed Chairs (since 2002), the collaborative research centre NanoBiC (since 2009) and the Beilstein Scholarship program (since 2011). In 2007 he was awarded his certificate of competence as project manager for his studies and thesis from the Studiengemeinschaft Darmstadt (a certified service provider). Since 2004 he coordinates the work of the STRENDA commission and promotes along with the commissioners the proposed standards of reporting enzyme data (www.strenda.org). These reporting standards have been adopted by, today, about 30 biochemical journals for their instructions for authors and are subject for the proposition of an electronic data capturing tool. Since 2011, Carsten co-ordinates the MIRAGE project which is concerned with the uniform reporting and representation of glycomics data in publications and databases. In 2014 Carsten became the head of the funding and conferences department which is also in charge of the foundation's public relationships.

Daniel Kolarich

2000	Diploma, Food Science and Biotechnology (University of Natural Resources and Applied Life Sciences, Vienna)
2004	PhD, (University of Natural Resources and Applied Life Sciences, Vienna), Topic: Mass spectrometry based glycoproteomic analysis of GMO food crops and allergens from plants and insects
2005 – 2007	Postdoc, (University of Natural Resources and Applied Life Sciences, Vienna); Research Interest in developing mass spectrometric tools for glycoconjugate characterisation
2007 – 2010	Post Doc, (Macquarie University, Sydney, Australia); Research interest in the role of glycosylation for sIgA and IgM function as well as bioinformatic tools for glycopeptide characterisation
Since 09/2010	Group Leader Glycoproteomics Group, (MPI of Colloids and Interfaces, Berlin), Research interest in development and automation of qualitative and quantitative LC-MS based glycoproteomics and glyomics to study the role of protein glycosylation in health and disease

Awards:
Sanofi-Aventis Price, Erwin Schrödinger Fellowship,
Förderpreis des Theodor Körner Fonds 2005

Editorial Board Memberships:
Journal of Molecular Recognition, *Proteomics*, *Glycoconjugate Journal*

Jung-Hsin Lin

also known as Jung-Hsing Lin (林榮信)

Current position and professional experience:

since 2010	Associate Research Fellow at the Research Center for Applied Science, Academia Sinica
2006 – 2010	Assistant Research Fellow at the Research Center for Applied Science, Academia Sinica
since 2004	Joint appointment at the Institute of Biomedical Sciences, Academia Sinica
since 2010	Associate Professorship at the School of Pharmacy, National Taiwan University

2003 – 2010	Assistant Professorship at the School of Pharmacy, National Taiwan University
2002 – 2003	Visiting scholar at the Computing Centre, Academia Sinica
2002	Visiting scholar at the Institute for Biological Information Processing, Forschungszentrum Jülich, Germany
2000 – 2002	Bioinformatics Specialist at the Howard Hughes Medical Institute, University of California San Diego, U.S.A.
2000	Postdoctoral researcher at the John von Neuman Institute for Computing, Forschungszentrum Jülich, Germany
1994 – 1996	Laboratory instructor at the Department of Physics, National Taiwan University
1993 – 1994	System and database administrator, Information Center, Army Headquarter, Republic of China (Taiwan)

Education:

1996 – 2000	Ph.D. In Biophysics, Institut für Festkörperforschung, Forschungszentrum Jülich and Institute of Physics, University of Duisburg, Germany
1990 – 1992	M.S., Graduate Institute of Physics, National Taiwan University
1986 – 1990	B.S., Department of Physics, National Taiwan University

Fields of Speciality:
Pharmacoinformatics, bioinformatics, computational biophysics, computational drug design, statistical physics, physical chemistry, theoretical pharmacy, high performance parallel computing, algorithms for optimization problems

Thisbe K. Lindhorst

is full professor at the Faculty of Mathematics and Natural Science of Christiana Albertina University of Kiel since 2000. She studied chemistry at the Universities of München and Münster, received her diploma in chemistry/biochemistry in 1988 and her Ph.D. in Organic Chemistry in 1991 at the University of Hamburg. After a postdoctoral stay at the University of British Columbia she worked on her habilitation and became Private Docent in 1998 at the University of Hamburg. In 1997 she was a Visiting Professor at the University of Ottawa in Canada. Since 2000 she holds a chair in Organic and Biological Chemistry in Kiel. Her most important awards are 'Förderpreis der Karl-Ziegler-Stiftung award', 'Chemiepreis der Akademie der Wissenschaften zu Göttingen award', and 'Carl-Duisberg-Gedächtnispreis award'. Her scientific interests are in the field of synthetic organic chemistry and in biological chemistry, especially in glycochemistry and glycobiology. Current research is focussed on the study of glycosylated surfaces and cellular adhesion. She is the author of over 100 publications and of the text book '*Essentials in Carbohydrate Chemistry and Biochemistry*'. She is an editor of the RSC Specialist Periodical Reports "*Carbohydrate Chemistry*" and of the Beilstein Thematic Issues "*Synthesis in the Glycosciences I and II*", amongst others. Her activities and professional responsibilities comprise of positions as head of chemistry department, elected board member of the Gesellschaft Deutscher Chemiker (GDCh), vice-chairwoman of the German Chemistry Gender Equality Initiative 'AKCC' within GDCh, and she is an elected member of the DFG Forschungsforum. She has invented a series of cultural events on value thinking for the university, named 'Wertedenken – Denkenswertes. Zur Zukunft der Universität', cf. http://www.wertedenken-denkenswertes.de. Thisbe is a mother of two almost grown-up children.

Hisashi Narimatsu

Director of Research Center for Medical Glycoscience, National Institute of Advanced Industrial Science and Technology (AIST)

Educational Training:

1974	Keio University, School of Medicine, M.D.
1979	Keio University, Postgraduate School of Medicine, Ph.D.

Positions and Honours:

1979, 4	Instructor, Department of Microbiology, Keio University School of Medicine
1983, 4	Postdoctoral Fellow, Laboratory of Immunology, National Institute of Allergy and Infectious Disease (NIAID), National Institutes of Health (NIH), Bethesda, Maryland, USA
1985, 5	Postdoctoral Fellow, Laboratory of Pathophysiology, National Cancer Institute (NCI), NIH, Bethesda, Maryland, USA. With Dr. Pradman Qasba.
1986, 4	Instructor, Department of Microbiology, Keio University School of Medicine
1986, 10	Associate Professor, Instructor, Department of Microbiology, Keio University School of Medicine
1991, 4	Professor, Division of Cell Biology, Institute of Life Science, Soka University
2000, 10	Principal Research Scientist, Group Leader of Gene Function Analysis, Institute of Molecular and Cell Biology, National Institute of Advanced Industrial Science and Technology (AIST), Japan
2002, 6	Group Leader of Glycogene Function Team, Deputy Director of Research Center for Glycoscience, National Institute of Advanced Industrial Science and Technology (AIST), Japan Professor, Tsukuba University School of Medicine (to present)
2006, 12 – present	Group Leader of Glycogene Function Team, Director of Research Center for Medical Glycoscience, National Institute of Advanced Industrial Science and Technology (AIST), Japan

2001, 4 – present Professor, School of Medicine Keio University
2011, 7 – present Advisory Professor, Shanghai Jiao Tong University

Milos V. Novotny

received his undergraduate education and a doctoral degree in biochemistry at the University of Brno, Czechoslovakia. His postdoctoral training in separation science and bioanalytical chemistry includes fellowships at the Czechoslovak Academy of Sciences, Royal Karolinska Institute (Sweden), and University of Houston.

Dr Novotny joined the IU Chemistry faculty in 1971, retiring from teaching and service in September 2011, but not from his research activities. He became James H. Rudy Professor in 1988, Distinguished Professor in 1999, and the Lilly Chemistry Alumni Chair in 2000. He has been known throughout the world for his pivotal role in developing modern chromatographic and electrophoretic methods of analysis.

His research interests include separation and structural analysis of biological molecules, proteomics and glycobiology, and chemical communication in mammals. Dr Novotny and his associates are known for structural identification of the first definitive mammalian pheromones. As a member of the Viking 1975 Science Team, Novotny designed the miniaturized GC column to search for organic molecules on the surface of Mars.

Milos Novotny has authored about 500 journal articles, reviews, books and patents. He has received more than 40 awards, medals and distinctions. His honors for chromatography include M.S. Tswett Medal (1984); American Chemical Society Award (1986); Keene P. Dimick Award (1990); Marcel J.E. Golay Award and Medal (1991); American Chemical Society Award in Separation Science and Technology (1992). For contributions to analytical chemistry, he received the 1988 ISCO Award in Biochemical Instrumentation; American Chemical Society Award in Chemical Instrumentation (1988), ANACHEM Award (1992); Pittsburgh Conference Analytical Chemistry Award (2000); and the EAS Award for Outstanding Achievements in the Fields of Analytical Chemistry (2001); Jan Weber Prize and Medal (2007); Ralph N. Adams Award in Bionalytical Chemistry (2008), and LC-GC Magazine (Europe) Lifetime Achievement Award in Chromatography (2012).

Overseas, he was honored through the Jan E. Purkynje Medal of the Czech Academy of Sciences; M.S. Tswett Memorial Medal (Russian Academy of Sciences); A.J.P. Martin Gold Medal and Theophilus Redwood Award (The Royal Society of Chemistry, Great Britain); Congreso Latinoamericano de Cromatografia Merit Medal (Argentina).

He is a recipient of honorary doctoral degrees from Uppsala University, Sweden (1991), Masaryk University, Czechoslovakia (1992), and Charles University, Czech Republic (2007).

Professor Novotny was named the 1994 Scientist of the Year by R&D Magazine. He is the member of two foreign academies: The Royal Society for Sciences (1994, Sweden) and The Learned Society of Czech Republic (2004).

At IU, he was a Distinguished Faculty Research Lecturer (1989), received the Distinguished Teaching and Mentoring Award of the University Graduate School (1997), the Distinguished Faculty Award of the College of Arts and Sciences (1999), and Tracy Sonneborn Award 2004.

Stefan Oscarson

got his PhD from the Department of Organic Chemistry at Stockholm University in 1985. He then worked one year in the research department of the Swedish Tobacco Company before returning to Stockholm University now as an Assistant Professor. He quickly built up a research group working on synthesis of biologically active oligosaccharides and glyco-conjugates and development of methodologies to facilitate these syntheses. In 1993 he was promoted to Associate Professor and in 2000 to Full Professor, all at the same department. In 2003 he accepted an offer to take over the Chair in Organic Chemistry at Gothenburg University and started building a new research group there working in the same research area. After a year he went back to his position at Stockholm University but kept his research group in Gothenburg until 2006. In 2006 he was appointed Professor of Chemical Biology at University College Dublin. Initially he spent time at both UCD and SU but since 2008 he is full-time at UCD where he is now supervising a group of around 13 researchers funded by grants from SFI, EU and NIH. He is an editor of *Carbohydrate Research* and the Swedish representative of the International and European Carbohydrate Organisation. He has supervised 26 PhD theses and published more than 160 peer-reviewed papers and 5 book chapters, all in the field of Glycoscience with special interests in glycoconjugate vaccines and carbohydrate-protein interactions involved in microbial adhesion and cell interactions.

Nicolle H. Packer

Nicki has had an extensive and varied career in biochemical research in both Chemistry and Biological Sciences. She was part of the team that established the Australian Proteome Analysis Facility (APAF) at Macquarie University in 1996 and left the University in 1999 to co-found Proteome Systems Limited, an Australian biotechnology company in which her group developed a platform of glycoanalytical technology and informatics tools.

For the past 6 years she is a Professor and Director of the Biomolecular Frontiers Research Centre at Macquarie University. Her research interests are in the structure and function of glycans, particularly in their role in cancer and microbial infection.

Nicki is involved in international efforts to compare the analytical protocols used in glycomics and was primary author of a NIH White Paper on recommendations for the glycoanalytical research area in 2008. She is currently involved in the international standardisation efforts in the field and continues to be invited to international discipline-specific workshops in US, Asia and Europe where the future requirements of the field are being addressed. Nicki has published extensively on glycomics research, is a senior editor of Proteomics, works closely with industry and is a Council member of HUPO. Her other achievements include producing 3 reasonably well-balanced children...

Vernon N. Reinhold

Education:

1959	B.S., Chemistry, University of New Hampshire
1961	M.S., Biochemistry, University of New Hampshire
1965	Ph.D., Biochemistry, University of Vermont

Research and professional Experience:

1968 – 1971	Helen-Hay Whitney Fellow, Massachusetts Inst. of Technology – Harvard University Medical School
1971 – 1973	Research Associate, DRR/NIH Mass Spec. Resource; MIT
1973 – 1975	Senior Investigator, Arthur D. Little, Inc. Cambridge
1976 – 1983	Lecturer, Harvard Medical School, Boston, MA
1984 – 1995	Lecturer in Biological Chemistry, Harvard University School of Medicine; Department of Biological Chem. & Pharm.; School of Public Health; Department of Nutrition.
1996 – 1998	Professor of Microbiology, Boston University School of Medicine, Boston, MA
1998 – present	Adjunct Professor of Microbiology, Boston University School of Medicine, Boston, MA
1998 – present	Res. Professor of Biochemistry, Department of Biochemistry & Molecular Biology, University of New Hampshire
1998 – present	Res. Professor of Chemistry, Department of Chemistry, University of New Hampshire, Durham, NH

Honours and professional services (selection):

1987 – 1997	Editorial Board, *Carbohydrate Chemistry*
1996 – 2003	Editorial Board, *Glycobiology*
1999	President, Society for Glycobiology

1999	Organizer 27th Annual Meeting of the Society for Glycobiology, San Francisco
2000 – present	Editorial Board, *Glycoconjugate Journal*
2001	Co-Chair ACS Northeast Regional Meeting, NERM

Pauline M. Rudd

obtained a BSc in Chemistry at the University of London and a PhD in Glycobiology at the Open University, UK. She was a Founding Scientist of Wessex Biochemicals (later Sigma London), Visiting Research Associate at The Scripps Research Institute, CA, Visiting Professor of Biochemistry at Shanghai Medical University PRC, Visiting Scientist at Ben Gurion University of the Negev, Israel and Erskine Visiting Fellow, Canterbury University, Christchurch, New Zealand. She is a Fellow of the Royal Society of Medicine, London and a Visiting Professor at St. George's Hospital, London and an Adjunct Professor at North Eastern University, Boston. She has more than 250 scientific publications and given over 300 lectures and seminars at international meetings. In 2010 she was awarded the James Gregory Medal and an Agilent Thought Leader award and in 2012 she received a Waters Global Innovation award.

Before moving her group to Dublin in 2006, Professor Rudd was a member of the Glycobiology Institute for 25 years.

Peter H. Seeberger

received his Vordiplom in 1989 from the Universität Erlangen-Nürnberg, where he studied chemistry as a Bavarian government fellow. In 1990 he moved as a Fulbright scholar to the University of Colorado where he earned his Ph.D. in biochemistry under the guidance of Marvin H. Caruthers in 1995. After a postdoctoral fellowship with Samuel J. Danishefsky at the Sloan-Kettering Institute for Cancer Research in New York City he became Assistant Professor at the Massachusetts Institute of Technology in January 1998 and was promoted to Firmenich Associate Professor of Chemistry with tenure in 2002. From June 2003 until January 2009 held the position of Professor for Organic Chemistry at the Swiss Federal Institute of Technology (ETH) in Zurich, Switzerland where he served as chair of the laboratory in 2008. In 2009 he assumed positions as Director at the Max-Planck Institute for Colloids and Surfaces in Potsdam and Professor at the Free University of Berlin. Since 2003 he serves as an Affiliate Professor at the Sanford-Burnham Institute in La Jolla, CA.

Professor Seeberger's research has been documented in over 250 articles in peer-reviewed journals, two books, fifteen issued patents and patent applications, more than 100 published abstracts and more than 500 invited lectures. Among other awards he received the Arthur C. Cope Young Scholar and Horace B. Isbell Awards from the American Chemical Society (2003), the Otto-Klung Weberbank Prize for Chemistry (2004), the Havinga Medal (2007), the Yoshimasa Hirata Gold Medal (2007), the Körber Prize for European Sciences (2007), the UCB-Ehrlich Award for Excellence in Medicinal Chemistry (2008), the Claude S. Hudson Award for Carbohydrate Chemistry from the American Chemical Society (2009) and the Tetrahedron Young Investigator Award for Medicinal and Bioorganic Chemistry (2010). In 2007 and 2008 he was selected among "The 100 Most Important Swiss" by the magazine "Schweizer Illustrierte". In 2011 he received the Hans Herloff Inhoffen-Medal.

Peter H. Seeberger is the Editor-in-Chief of the *Beilstein Journal of Organic Chemistry* (2011 – present), was the Editor of the *Journal of Carbohydrate Chemistry* (2003 – 2010) and serves on the editorial advisory boards of many other journals. He is a founding member of the board of the Tesfa-Ilg "Hope for Africa" Foundation that aims at improving health care in Ethiopia in particular by providing access to malaria vaccines and HIV treatments. He is a consultant and serves on the scientific advisory board of several companies. In 2006 he served as president of the Swiss Academy of Natural Sciences.

The research in professor Seeberger's laboratory has resulted in two spin-off companies: Ancora Pharmaceuticals (founded in 2002, Medford, USA) that is currently developing a promising malaria vaccine candidate in late preclinical trials as well as several other therapeutics based on carbohydrates and i2chem (2007, Cambridge, USA) that is commercializing microreactors for chemical applications.

Jürgen Seibel

born in 1971, studied chemistry at the Georg August University of Göttingen and received his diploma in 1998. During his PhD thesis under the guidance of Professor Dr. Dr. h. c. Lutz F. Tietze, he was involved in the synthesis of Vitamin E. In 2000, he moved to the University of Oxford to work with Professor Dr. Chris Schofield in the Dyson Perrins Laboratory (Department of Organic Chemistry) on serine proteases and Hypoxia-inducible factors (HIF) on the mechanisms and as therapeutic targets. In 2002, he came back to Germany starting his independent research at the University of Braunschweig. In 2006 he finished his habilitation. In 2007 he moved to Helmholtz Centre for Infection Research (HZI). In 2009 he became professor at the University of Würzburg. His present research interests are focused on Bioorganic Chemistry and Chemical Biology including the development of chemical and enzymatic syntheses, biocatalysis, protein engineering, drug delivery and glycosciences. In 2008 he was awarded with the Jochen-Block prize by the Dechema (Society for Chemical Engineering and Biotechnology). In 2012 he was honored as DuPont Young Professor.

Index of Authors

Index